Competition Models in Population Biology

PAUL WALTMAN
Emory University

SOCIETY for INDUSTRIAL and
APPLIED MATHEMATICS • 1983

PHILADELPHIA, PENNSYLVANIA 19103

Copyright © 1983 by Society for Industrial and Applied Mathematics.

Library of Congress Catalog Card Number: 83-50665.
ISBN: 0-89871-188-6

Contents

Preface v

Chapter 1
SINGLE POPULATION GROWTH MODELS
   1. Introduction 1
   2. Logistic growth 1
   3. The simple chemostat 6

Chapter 2
INTERACTING POPULATIONS
   1. Introduction 11
   2. The Lotka-Volterra competition model 14
   3. Competition in the chemostat 18
   4. A three-level food chain 25
   5. Bifurcation from a limit cycle 30
   6. Two predators competing for a renewable resource 34
   7. Two predators feeding on a prey growing in a chemostat 41

Chapter 3
SOME DETERMINISTIC PROBLEMS IN GENETICS
   1. Introduction 47
   2. Predator influence on the growth of a population with three genotypes   51
   3. A model of cystic fibrosis 58
   4. A parental selection problem 65

References 75

Preface

These lecture notes are from a series of lectures given at a regional conference at Logan, Utah, in July of 1982. The audience had a wide variety of backgrounds, and principal groups included those interested in biology or forestry, those interested in ordinary differential equations but not particularly in biology, and those whose major interest was in mathematical biology. The lectures began on familiar territory to ease some mathematical anxiety and proceeded to new material. The single theme was the use of relatively elementary ideas of dynamical systems to answer questions of a biological nature, in particular questions as to the nature of the eventual (asymptotic) behavior of populations given a relatively few hypotheses as to the nature of their growth and their interactions.

The success of the conference depended very much on the enthusiasm of the participants, the extremely strong supporting lectures and the work of the organizing committee, in particular the untiring efforts of Bob Gunderson and Bob McKelvey. The very lovely surroundings of the campus of Utah State University and the assistance of the Mathematics Faculty contributed a great deal to the enjoyment of the conference.

Finally, I would like to acknowledge, with thanks, the efforts of K. Beck, H. Freedman, J. A. Gatica and S. B. Hsu, who read and commented on portions of the original manuscript.

CHAPTER 1

Single Population Growth Models

1. Introduction. During the course of these lectures an attempt will be made to describe the interactions of populations in the language of mathematics. The term population is used in the wide sense; it may denote the usual population of individual organisms or it may be used for molecules, cells, etc. Before considering interacting populations, however, it is necessary to consider the growth of a single population. Here we will focus on two models, one commonly called logistic growth and one which describes growth in a piece of laboratory apparatus, the chemostat.

Consider first a population of organisms. We wish to describe how the population changes with time. The basic concept is that of the intrinsic rate of growth—the per capita growth rate. Let $x(t)$ denote the number of members of the population at time t. Then the intrinsic growth rate is defined by $x'(t)/x(t)$, where $x'(t)$ denotes the derivative d/dt. A model of population growth specifies how this quantity changes with the size of the population. Symbolically one writes

$$\frac{x'(t)}{x(t)} = \text{MODEL}.$$

Below, "MODEL" is specified in two different ways.

2. Logistic growth. The simplest possible model would be to make the intrinsic growth rate a constant, to specify

$$\frac{x'(t)}{x(t)} = r, \qquad r \geq 0.$$

This is just the differential equation

$$x'(t) - rx(t) = 0$$

whose solutions are given by $x(t) = x(0) \exp(rt)$. This model is usually associated with the name of Malthus, and if r is not zero, predicts unbounded growth. The usual interpretation of r is that it is birth rate minus death rate, $r = b - d$, so that these two quantities must be in balance to prevent unlimited growth. Such a model does not meet with observations, and we discard it. The next simplest model is to assume that the intrinsic rate of growth is given as a first degree polynomial,

$$\frac{x'(t)}{x(t)} = ax + b.$$

This is usually written

(2.1) $$x'(t) = rx\left(1 - \frac{x}{K}\right)$$

where $K > 0$ is called the carrying capacity and $r > 0$ is called the maximal growth rate. This equation is known as the logistic equation and is associated with the name of Verhulst. The variables separate and it can be solved explicitly to yield that if $x(0) > 0$, then $\lim_{t \to \infty} x(t) = K$. Interesting reading on the history of this equation and examples of its applicability can be found in the book by Hutchinson [40]. Solutions of this equation fit a number of biological populations and a few examples from this book are discussed below. Before discussing the applications, we note two of Hutchinson's cautions on answering the question, "Does the logistic equation fit the data?". First of all it is too much to expect a single equation to express growth in a limited universe. Secondly, if a great deal is known about an organism, then constructing an equation that is fit by careful experiments tells us nothing—the organism merely acts as an analog computer. Armed with these cautions we proceed to look at some data.

Since the mathematical assumptions for population growth are best met for simple organisms, it is not surprising that the best fit between model and data occurs for microorganisms. Figure 2.1 shows the fit of a solution of the logistic equation to data for the growth of Escherichia coli. (All of the figures in this section are from the book of Hutchinson [40] and are reproduced by permission of the publisher, Yale University Press.) Figure 2.2 shows the growth of a population of Paramecium caudatum fitted with a solution of the logistic equation. More interesting, perhaps, is Fig. 2.3 which shows the growth of the collared turtledove in Great Britain. The

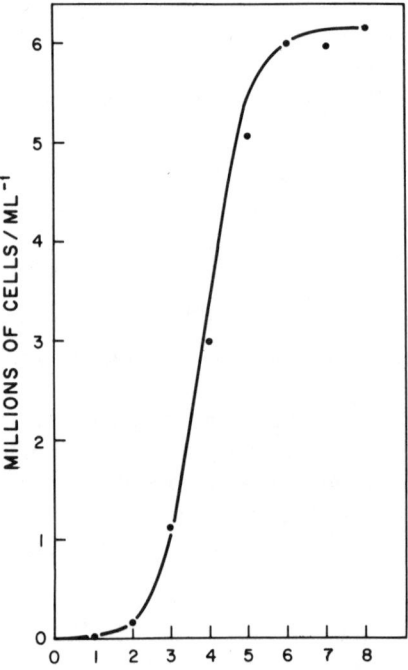

FIG. 2.1. *Logistic curve fitted to the growth of Escherichia coli. From Hutchinson* [40], *data of McKendrick and Kesava Pai.* [Copyright 1978, Yale University Press. Used by permission.]

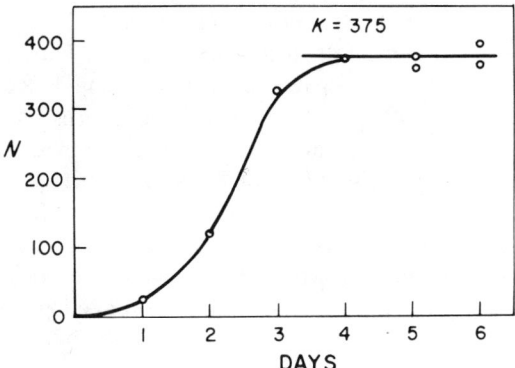

FIG. 2.2. *Logistic curve fitted to the growth of Paramecium caudatum. From Hutchinson* [40], *data of Gause.* [Copyright 1978, Yale University Press. Used by permission.]

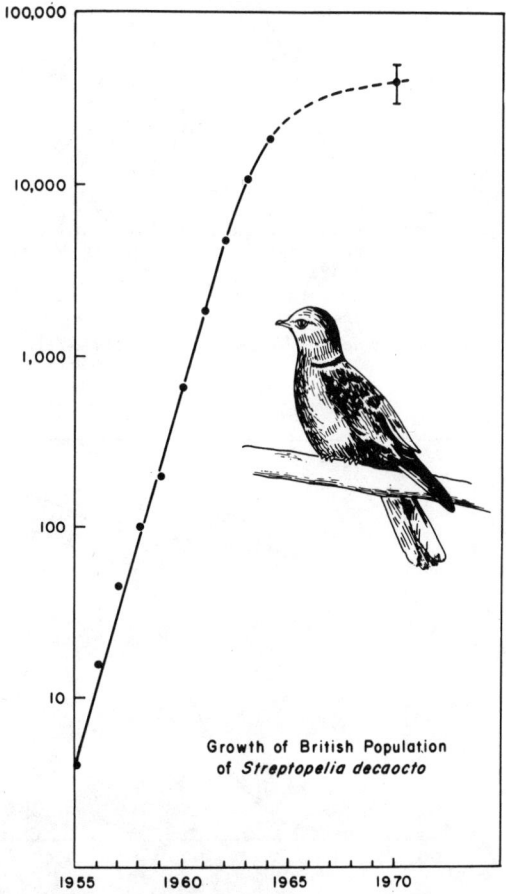

FIG. 2.3. *Logistic curve fitted to the estimation of population of collared turtledove in Great Britain. From Hutchinson* [40]. [Copyright 1978, Yale University Press. Used by permission.]

bird was first noticed in Great Britain (by birdwatchers) in about 1955; Fig. 2.3 shows the agreement between data and the logistic curve. The data are less reliable at the end because once the bird became less rare it was no longer of interest to the birdwatchers. Figure 2.4 shows a similar plot of honeybees (Apis mellifera) near Baltimore. In this case carrying capacity is determined by hive size, so two different colonies grow to almost identical values of K. Now that it seems that the logistic equation might fit everything, it is appropriate to show something that doesn't fit. Figure 2.5 shows a logistic curve fit to the U.S. census data up to 1940. It clearly underestimates the population growth. Finally to show that logistic growth may (unexpectedly) fit almost anything, Fig. 2.6 shows a fit of a logistic curve to a plot of

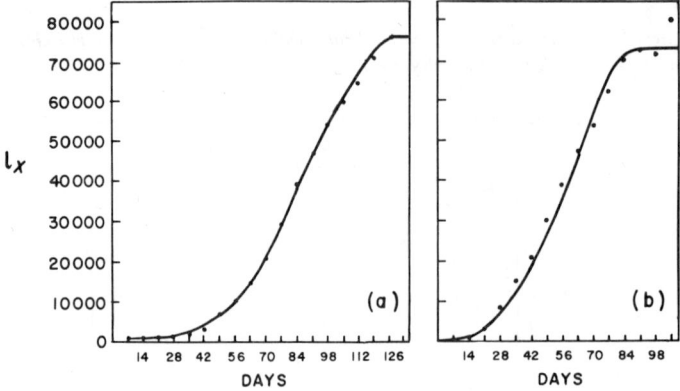

FIG. 2.4. *Logistic curve fitted to the growth of colonies of honeybees near Baltimore. From Hutchinson* [40], *data from Bodenheimer and Nolan.* [Copyright 1978, Yale University Press. Used by permission.]

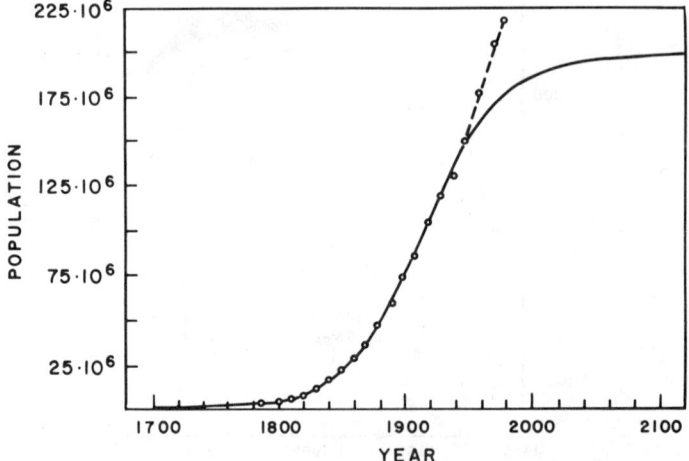

FIG. 2.5. *Logistic curve fitted to U.S. census. From Hutchinson* [40], *original from Pearl, Reed and Kish.* [Copyright 1978, Yale University Press. Used by permission.]

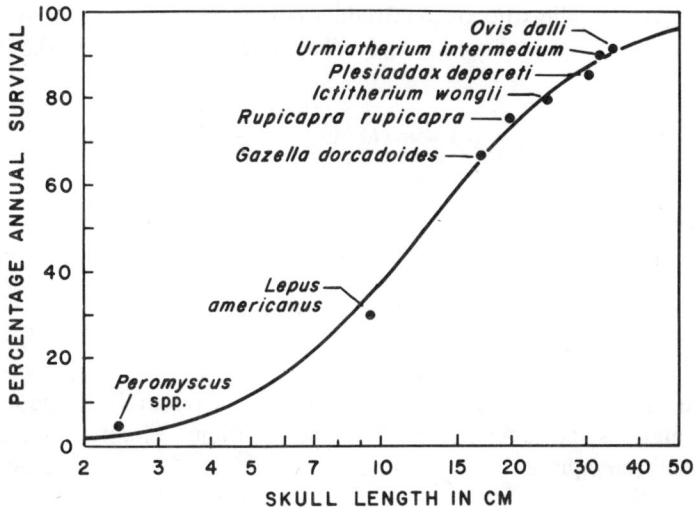

FIG. 2.6. *Logistic curve fitted to the annual percentage of survival for mammals plotted against skull length. From Hutchinson* [40], *citing Kurten.* [Copyright 1978, Yale University Press. Used by permission.]

skull length against the percentage of annual survival—the larger one's skull the better one's chances of survival to the next year, with a leveling off appropriate for a solution of the logistic equation.

What conclusions should be drawn from this evidence? First of all, the logistic equation is not irrelevant for a description of biological growth—it fits microorganisms and certain other situations well. On the other hand, it is not appropriate for all types of population growth. After an experiment, data may be fit by a logistic curve, but it offers no explanation and no advance prediction. Moreover, it seems appropriate when there is a clearly identifiable carrying capacity.

There are alternatives to the logistic equation. The oldest, perhaps, is due to Gompertz, whose equation takes the form

$$x'(t) = rx \ln\left(\frac{K}{x}\right).$$

The solutions of this equation share the essential properties of the logistic equation—for small x the derivative is large and in the time limit (limit as t becomes infinite) the solution takes the value K. One can view both of these equations as generating three-parameter curve families, the parameters being r, K and the initial condition. Many other three-parameter curve families might do equally well.

There are several modifications of the logistic equation which introduce an additional parameter to make the equation more realistic. Gilpin and Ayala [26] proposed

$$x'(t) = rx\left[1 - \left(\frac{x}{K}\right)^\alpha\right].$$

Another popular modification is to introduce delays into the equation to account for a time lag between nutrient uptake and reproduction. The simplest takes the form

$$x'(t) = rx(t)\left[1 - \frac{x(t-s)}{K}\right].$$

The delays can also be distributed, yielding an equation of the form

$$x'(t) = rx(t)\left[1 + \int_0^t f(y)x(t-y)dy\right].$$

These two equations have motivated a great deal of research on the existence of periodic solutions for functional differential equations. Finally, one can write a general nonlinear equation

$$x'(t) = x[g(x) - d(x)]$$

and place hypotheses on g and d which reflect the properties of logistic growth.

3. The simple chemostat. The chemostat is a piece of laboratory apparatus used for culturing microorganisms. It has a constant nutrient source, containing all nutrients needed by the microorganism in abundance except one. The nutrient is pumped at a constant rate into a culture vessel. Constant volume is maintained in the culture vessel by allowing an overflow or by pumping the contents of the vessel out at the same rate that the nutrient is pumped in. The output of the culture vessel is collected in a receptacle. The culture vessel is charged with a quantity of a given type of microorganism and the collection vessel then contains both organism and nutrient. This provides a continuous supply of microorganisms, and hence the name continuous culture is used to contrast the process with the more familiar batch culture technique. For ecological purposes the chemostat is the laboratory realization of a very simple lake; the importance of the chemostat as an experimental vehicle is well documented, for example in the survey articles [21], [22], [40], [60], [61]. It is also of interest in chemical engineering where it is a simplified model of the wastewater treatment process [54].

For purposes of mathematical modeling, the important properties are that the input nutrient has constant concentration, that the input flow rate is constant (both of these are under the control of the experimenter), that the culture vessel has constant volume, that the mixture is well stirred, that the temperature is held constant, and that the flow rate is high enough that metabolic products wash out "sufficiently fast" so that they have no effect. Implicitly it is assumed that no mutations occur over the length of the experiment—this is not always the case (and in fact chemostats are used in genetic experiments), so it is necessary at the end of an experiment to ascertain that the organism finally cultured is in fact the one with which the culture vessel was charged.

Let S denote the nutrient concentration (S stands for substrate, suggested by the eventual form of the equations), $S^{(0)}$, the input concentration, and x, the concentration of microorganisms. If the nutrient were not being consumed then a model would be

$$S' = (S^{(0)} - S)D,$$

which expresses the fact that the rate of change of a concentration is proportional to the difference between the incoming concentration and the present concentration of the vessel. With consumption, then, it takes the form

$$S' = (S^{(0)} - S)D - [\text{CONSUMPTION}].$$

For the microorganism one writes

$$[\text{RATE OF CHANGE}] = [\text{GROWTH}] - [\text{WASHOUT}].$$

The growth can be expressed in terms of the nutrient consumed by

$$[\text{GROWTH}] = \gamma[\text{CONSUMPTION}]$$

where γ is a yield constant expressed as

$$\gamma = \frac{\text{organism formed}}{\text{substrate used}}.$$

There remains only to postulate how the resource is consumed. This was done at least as far back as Monod [51] who, based on experimental evidence, used

$$\frac{mxS}{a+S},$$

where $m > 0$ is called the maximal growth rate and $a > 0$ is called the Michaelis–Menten constant (this last term coming from enzyme kinetics). In other contexts, this is an example of a Holling type II functional response [33]. The constant a is sometimes called the half saturation constant, as it is the concentration at which the per capita growth rate achieves $m/2$ or half the maximal rate. The equations take the form

$$S' = (S^{(0)} - S)D - \frac{mxS}{(a+S)},$$

$$x' = \frac{mxS}{a+S} - Dx, \qquad S(0) \geq 0, \quad x(0) > 0.$$

For the mathematical analysis it is easier to work with a nondimensional form. Accordingly, one can scale all of the concentrations by the input concentration and in addition scale the microorganism value by the yield constant (note that it is nondimensional already). m can be scaled by D, and time, by $1/D$. This produces the system

(3.1)
$$S' = 1 - S - \frac{mxS}{a + S},$$
$$x' = \frac{mxS}{a + S} - x, \qquad S(0) \geq 0, \quad x(0) > 0.$$

Before beginning the analysis of this system, some terminology and some facts from the theory of dynamical systems will be introduced. This procedure will be followed throughout these notes; concepts from the theory of ordinary differential equations will be introduced as needed. Although not absolutely necessary at this point, it is convenient to begin here. Consider an autonomous differential equation

(3.2)
$$x'(t) = f(x(t))$$

where f maps n-dimensional Euclidian space into itself. A convenient way to view solutions of (3.2) is in "phase space"—to view $x(t)$ as a parametric curve in Euclidean n-space. A periodic solution of (3.2), for example, would correspond to a closed curve. The curve given by a solution of (3.2) in phase space is called a *trajectory* or an *orbit*. It is convenient to divide a trajectory into two parts. The set of points corresponding to $t > 0$ is called the positive semiorbit and, to $t < 0$, the negative semiorbit. A set B is said to be *positively invariant* if the positive semiorbit of a trajectory is in B if the point corresponding to $t = 0$ is in B. *Negatively invariant* is defined similarly but with $t < 0$. A set B is *invariant* if it is positively and negatively invariant. The positive cone (the first quadrant) is positively invariant for trajectories of the system (3.1). Let t_n be a sequence of points such that $t_n \to \infty$ as $n \to \infty$. (Such a sequence is called an *extensive sequence*.) P is said to be an *omega limit point* of a trajectory of (3.2) if there exists an extensive sequence t_n such that $x(t_n)$ converges to P as $n \to \infty$. The set of all such points is called the *omega limit set* of a trajectory. An important fact is that *every bounded trajectory is asymptotic to its omega limit set*. A second important property is that if P is in the omega limit set of a trajectory, the closure of the orbit through P is also in the omega limit set. A critical point of the system (3.2) is a point a such that $f(a) = 0$. This constant value is a solution of the differential equation—terms other than critical point which are sometimes used are stationary point and equilibrium point. If the system (3.2) is two-dimensional then the omega limit set of bounded trajectories can only consist of critical points, trajectories connecting critical points and periodic orbits.

To analyze this equation we take a roundabout approach in order to illustrate the ideas which will appear later and to note the connection with the logistic equation. Let $z(t) = 1 - x(t) - S(t)$ where $S(t)$ and $x(t)$ satisfy (3.1). Then (3.1) is equivalent to

(3.3)
$$z' = -z,$$
$$x' = x\left[\frac{m(1 - x - z)}{1 + a - x - z} - 1\right], \qquad z(0) > 0, \quad x(0) > 0.$$

It then follows that $z(t) = z(0) \exp(-t)$, or that, as $t \to \infty$, $z(t) \to 0$ (exponentially). Thus the omega limit set of any trajectory of (3.3) lies on the line $z = 0$. Since every trajectory is asymptotic to its omega limit set, it is only necessary to study the trajectories there. This argument allows us to make the substitution $z = 0$ to obtain the scalar equation

(3.4) $$x' = x\left[\frac{m(1-x)}{1+a-x} - 1\right].$$

The proof of the following theorem is made with (3.3) and (3.4) although the theorem is stated for our original system (3.1). The proof is somewhat indirect in order to exhibit the similarity of the logistic equation (2.1) and the chemostat system (3.3).

THEOREM 3.1. *If $m \leq 1$ or if $m > 1$ and $\lambda = a/(m-1) \geq 1$, then*

$$\lim_{t \to \infty} x(t) = 0$$

for every solution of (3.1). *If $m > 1$ and $\lambda < 1$, then*

$$\lim_{t \to \infty} x(t) = 1 - \lambda, \quad \lim_{t \to \infty} S(t) = \lambda$$

for every solution of (3.1).

Proof. Let $x(t)$ be an arbitrary solution of (3.4). Since $z(t) = 0$, one can assume without loss of generality that $x(t) < 1$. Suppose $x(0) > 0$. If $m \leq 1$, then $x' \leq -ax/(1+a)$ and the conclusion of the theorem follows. Define

$$w = \int_0^t \frac{du}{1+a-x(u)}.$$

Note that $1/a > w' > 1/(1+a)$. Make a (solution-dependent) change of time scale from t to w. Then $x(t)$ is one of the trajectories of the logistic equation

(3.5) $$\frac{dx}{dw} = (m-1)x(1-\lambda-x)$$

where in (2.1) $r = (m-1)/(1-\lambda)$ and $K = 1 - \lambda$. If $m > 1$ and $\lambda < 1$ all trajectories of this equation with positive initial conditions tend to $1 - \lambda$ as w tends to infinity. If $m \leq 1$ or if $m > 1$ and $\lambda \geq 1$, then all solutions of (3.5) tend to zero as w tends to infinity. Since $t \to \infty$ if and only if $w \to \infty$, the same statements are true for $x(t)$ as t tends to infinity.

One small point remains. Could it be the case that $m > 1$ and $\lambda < 1$ and the orbit of (3.3) contains $(0, 0)$ in its omega limit set? (Equivalently, could $\liminf_{t \to \infty} x(t) = 0$?) Since $\lambda < 1$ there is an $\varepsilon > 0$ with $\varepsilon < 1 - \lambda$. Then no trajectory of (3.3) may cross the line $z = \varepsilon$ more than once. For $z(t) \leq \varepsilon$, $x'(t) > 0$, so $x(t)$ comes no closer to zero than its value when the trajectory of (3.3) crosses $z = \varepsilon$. This justifies using (3.4) instead of (3.3).

CHAPTER 2

Interacting Populations

1. Introduction. The growth models considered in Chapter 1 consisted of a single population. (The chemostat actually began with two but was reduced to a mathematical problem involving only one.) We will build on these two growth models and study more interesting ecological phenomena by adding populations to the model. Specifying the nature of the interactions constitutes a biological hypothesis, and we will restrict our considerations to two specifications—mass action and Michaelis–Menten. Mass action is a term borrowed from chemistry, and states that a reaction between two substances proceeds at a rate proportional to the amount (or concentration) of the interacting chemicals. The same hypothesis may be made for populations. The Michaelis–Menten hypothesis has already been introduced, and its key feature is that the rate is damped or limited as one of the concentrations increases. Increasing the number of populations in the model raises the mathematical complexity and hence new techniques are required. Some of these (for example, the necessary bifurcation theory and the Floquet theory) will be introduced as needed, but we begin this chapter by reviewing some facts from the theory of plane autonomous systems.

Consider a system of differential equations of the form

$$(1.1) \qquad x' = f(x, y), \qquad y' = g(x, y)$$

where f and g are continuously differentiable functions from R^2 to R. It will also be assumed that solutions continue for all time (or at least for all positive time). Neither of these hypotheses is restrictive, since the models will usually be real analytic and all solutions will be bounded (at least in the future, i.e., for $t > 0$). The phase space discussed in Chapter 1 is just the (x, y)-plane. An important property of the phase space is that through each point there is a unique orbit, with the consequence that a trajectory may not cross itself, and that a trajectory which intersects itself (a simple closed curve) is the orbit of a periodic solution. This makes the use of geometric and topological ideas profitable in the analysis of systems of the form (1.1).

The idea of omega limit set and the fact that a bounded trajectory is asymptotic (as $t \to \infty$) to its omega limit set were introduced in Chapter 1. There is an analogous concept of *alpha limit set* if one replaces t_n in the definition of omega limit point by $-t_n$. In fact, if one replaces t by $-t$ in (1.1), trajectories—the curves—remain the same but alpha and omega limit sets are interchanged (if they both exist). If a trajectory is bounded in the future then the omega limit set is nonempty, compact, connected and invariant. As noted in Chapter 1 the omega limit set of a system such as (1.1) can consist of critical points, critical points and orbits joining them, or closed curves. Another way to view trajectories of (1.1) is through the solution map. Fix a time τ and define a mapping of the plane into itself by defining $T(x^*, y^*)$ to be the point $(x(\tau), y(\tau))$ where $(x(t), y(t))$ is the solution of (1.1) with initial conditions $x(0) = x^*$ and $y(0) = y^*$. The trajectory then is just all the images of the solution map as t ranges over all real numbers.

A critical point, (x^*, y^*), of (1.1) satisfies

$$f(x^*, y^*) = 0, \qquad g(x^*, y^*) = 0.$$

A critical point (x^*, y^*) is said to be *stable* if for every $\varepsilon > 0$ there is a $\delta > 0$, such that if $(x(0), y(0))$ is within δ of (x^*, y^*) then $(x(t), y(t))$ is within ε of (x^*, y^*) for all $t > 0$. A critical point (x^*, y^*) is said to be *asymptotically stable* if it is stable and if the above $(x(t), y(t))$ tends to (x^*, y^*) as $t \to \infty$.

(1.1) may be rewritten

$$x' = a(x - x^*) + b(y - y^*) + e_1(x - x^*, y - y^*),$$
$$y' = c(x - x^*) + d(y - y^*) + e_2(x - x^*, y - y^*)$$

where

$$A = \begin{bmatrix} a & b \\ c & d \end{bmatrix}$$

is the Jacobian of f and g evaluated at the critical point, and e_i, $i = 1, 2$, are continuously differentiable. A is called the variational matrix of (1.1) at (x^*, y^*). Moving (x^*, y^*) to the origin yields a system of the form

(1.2) $\qquad x' = ax + by + e_1(x, y), \qquad y' = cx + dy + e_2(x, y).$

The behavior of trajectories of (1.2) near the origin is related to the behavior of trajectories of the linearization

(1.3) $\qquad\qquad\qquad x' = ax + by, \qquad y' = cx + dy.$

The next two theorems make precise the portion of this relationship that will be used in the following sections. All of the theorems presented in this section are special cases of classical results which may be found in most differential equations texts, for example, in [3] or [16].

THEOREM 1.1. *If the eigenvalues of A have negative real parts, then $(0, 0)$ is a (locally) asymptotically stable critical point of* (1.2).

The trajectories of (1.2) in a region G_1 are said to have the *same qualitative structure* as trajectories of (1.3) in a region G_2 if there is a homeomorphism of G_1 onto G_2 such that it and its inverse map trajectories into trajectories.

THEOREM 1.2. *If no eigenvalue of A has zero real part, then* (1.2) *and* (1.3) *have the same qualitative structure.*

Much more precise geometric information can be obtained about the behavior of trajectories of (1.2) in terms of the behavior of the linearization (1.3) when no eigenvalue of A has zero real part. For example, if the eigenvalues of A are of opposite sign, then the origin is said to be a saddle point for (1.3). There will exist two sets, S and U, of dimension one, which intersect only at the origin. S is positively invariant and such that the origin is the omega limit set of any trajectory with initial conditions on S. U is negatively invariant and the origin is the alpha limit set of any trajectory with initial conditions on U. S is called the *stable manifold* and U, the *unstable manifold*. When the eigenvalues of A are of opposite sign, these same sets U and S exist for the system (1.2) in a neighborhood of the critical point. In particular, there will be exactly two trajectories which tend to the critical point as $t \to \infty$ and exactly two trajectories which tend to the critical point as $t \to -\infty$. No other trajectory may have this critical point as its alpha or omega limit set. It will frequently happen, in the models which follow, that there are saddle points on the boundary of a quadrant and the stable manifolds lie on an axis. It can be shown that this will preclude the critical point from being the omega limit set of any trajectory with initial conditions in the interior of the quadrant, unless the omega limit set contains an orbit on the stable and unstable manifolds. Note also that if an asymptotically stable critical point is in the omega limit set of a trajectory, then it is the omega limit set. Similarly, a repeller (both eigenvalues of A have positive real part) cannot be in an omega limit set of a trajectory of (1.2). There are obvious generalizations to n-dimensional systems such as (3.2) of Chapter 1. The next theorems apply only to two-dimensional systems.

The following theorem provides for the existence of periodic solutions of (1.1).

THEOREM 1.3 (Poincaré–Bendixson). *Let Γ be a trajectory of (1.1) which remains in a closed and bounded region for $t > 0$ and let Ω be its omega limit set. If Ω contains no critical points, then either Γ is a periodic trajectory (and $\Omega = \Gamma$) or Ω is a periodic trajectory.*

Closed orbits which are the alpha or omega limit set of other orbits are called limit cycles. The following theorem is useful for eliminating the possibility of "unwanted" closed orbits in an open (invariant) quadrant when all of the critical points lie on the boundary.

THEOREM 1.4. *Every closed trajectory of (1.1) has a critical point in its interior.*

The elimination of closed orbits when there is an appropriate critical point is more difficult and is an area in need of more research. The following is one of the few theorems in this direction.

THEOREM 1.5 (Dulac criterion). *If $\beta(x, y)$ is a continuously differentiable function such that*

$$(\beta(x, y)f(x, y))_x + (\beta(x, y)g(x, y))_y$$

does not change sign in a simply connected region G, then (1.1) has no closed orbits in G.

When $\beta = 1$, this result is known as Bendixson's negative criterion.

A general method of attack for several of the problems considered is the following recipe:

i) Locate the critical points and linearize about each one.
ii) Determine the local stability and find that there will be only one asymptotically stable critical point.
iii) Show that all trajectories are bounded for $t > 0$.
iv) Eliminate the possibility of limit cycles and saddle connections.
v) Conclude that the omega limit of all trajectories is a single critical point.

If ii) is not true then one must work harder to determine the eventual asymptotic behavior. In some models, limit cycles will exist and further analysis is required.

The emphasis in the models discussed in Chapters 2 and 3 is on models of competition. Many other kinds of interactions between populations are possible, and there is a vast literature on the subject. The interested reader might begin with some of the following books: Christiansen and Fenchel [15], Freedman [23], Hoppensteadt [34], Hutchinson [40], May [48], Maynard Smith [49].

2. The Lotka–Volterra competition model. The first model of interacting populations to be considered is a classical one, usually associated with the names of Lotka and Volterra, a model which has now found its way into many mathematics textbooks. It is included here to illustrate the use of the theorems on dynamical systems outlined in §1. If two populations were growing logistically and not affecting each other, their growth could be described by two (uncoupled) logistic equations:

$$x' = r_1 x \left(1 - \frac{x}{K_1}\right),$$

$$y' = r_2 y \left(1 - \frac{y}{K_2}\right), \quad x(0) = x_0 > 0, \quad y(0) = y_0 > 0.$$

Assume, however, that the carrying capacity is a shared resource—each population competes for the resource and thereby interferes with the other. Then the presence of each reduces the intrinsic rate of growth of the other. This can be expressed by

$$x' = r_1 x \left(1 - \frac{x}{K_1} - \lambda_1 y\right),$$

$$y' = r_2 y \left(1 - \frac{y}{K_2} - \lambda_2 x\right), \quad x(0) = x_0 > 0, \quad y(0) = y_0 > 0$$

where r_i, K_i, and λ_i, $i = 1, 2$, are positive quantities. It is convenient to change to nondimensional variables by measuring x in units of K_1, y in units of K_2 and time in units of $1/r_1$. The equations then take the form

(2.1) $\quad\quad x' = x(1 - x - \lambda_1 y), \quad y' = ry(1 - y - \lambda_2 y).$

The new λ_i is the old λ_i times K_i and r is the quotient r_2/r_1. (2.1) is called the Lotka–Volterra competition model, and it is this equation that will be analyzed.

First it is necessary to settle some minor concerns about the behavior of trajectories. That the positive quadrant is positively invariant can be seen by representing a solution in integral form:

$$x(t) = x(0) \exp\left[\int_0^t [1 - x(s) - \lambda_1 y(s)] ds\right],$$

$$y(t) = y(0) \exp\left[r \int_0^t [1 - y(s) - \lambda_2 x(s)] ds\right].$$

Thus if the initial conditions are positive, the components of the solution are positive for all finite time t. A solution pair of (2.1) also satisfies the differential inequality

$$x' \leq x(1 - x), \qquad y' \leq ry(1 - y).$$

x may be compared with the solution of

$$z' = z(1 - z), \qquad z(0) = x(0)$$

and y may be compared with

$$z' = rz(1 - z), \qquad z(0) = y(0)$$

to show that both $x(t)$ and $y(t)$ are bounded.

There are four (possible) critical points which we label
E_0:(0, 0),
E_1:(0, 1),
E_2:(1, 0),
E_3: the solution—if it exists—of

$$x + \lambda_1 y = 1, \qquad \lambda_2 x + y = 1.$$

The linearization of (2.1) takes the form

$$A = \begin{bmatrix} 1 - 2x^* - \lambda_1 y^* & -\lambda_1 x^* \\ -r\lambda_2 y^* & r(1 - 2y^* - \lambda_2 x^*) \end{bmatrix}$$

where (x^*, y^*) is one of the above critical points.

At E_0, the matrix A becomes

$$\begin{bmatrix} 1 & 0 \\ 0 & r \end{bmatrix},$$

and so both eigenvalues are positive and the origin is a repeller. At E_1, A takes the form

$$\begin{bmatrix} 1 - \lambda_1 & 0 \\ -r\lambda_2 & -r \end{bmatrix},$$

so that one eigenvalue is negative and one has the sign of $1 - \lambda_1$. Thus E_1 either is a saddle point or is asymptotically stable. At E_2, A is of the form

$$\begin{bmatrix} -1 & -\lambda_1 \\ 0 & r(1 - \lambda_2) \end{bmatrix}$$

so one eigenvalue is again negative and one has the sign of $1 - \lambda_2$. Hence it either is a saddle point or is asymptotically stable. Finally, to determine the stability of E_3, one solves

(2.2) $$x + \lambda_1 y = 1, \qquad \lambda_2 x + y = 1$$

to get

$$x^* = \frac{1 - \lambda_1}{1 - \lambda_1 \lambda_2}, \qquad y^* = \frac{1 - \lambda_2}{1 - \lambda_1 \lambda_2}.$$

E_3 is meaningful only if $x^* > 0$ and $y^* > 0$. If we view the equations as describing two lines in the phase plane, there are four cases to be analyzed, depending on the relative position of the intercepts. These are shown in Fig. 2.1.

In cases a and b, the lines do not intersect in the open positive quadrant, so there are no critical points there. Hence there can be no periodic solutions. In case a, $\lambda_1 - 1 < 0$ and $\lambda_2 - 1 > 0$, so E_1 is a saddle point and E_2 is an asymptotically stable attractor. If E_i, $i = 1, 2$, is a saddle point, its stable manifold lies along an axis, so it cannot be in the omega limit set of any orbit with initial conditions interior to the positive quadrant. Thus, from the Poincaré–Bendixson theorem, the omega limit set of any trajectory of (2.1) is the critical point E_1; $(0, 1)$ is a globally (with respect to the open positive quadrant) asymptotically stable attractor. The biological implication is that y wins the competition and x loses (x becomes "extinct"). This is an example of competitive exclusion—only one competitor survives. In case b the roles of E_1 and E_2 are reversed, but still only one competitor survives.

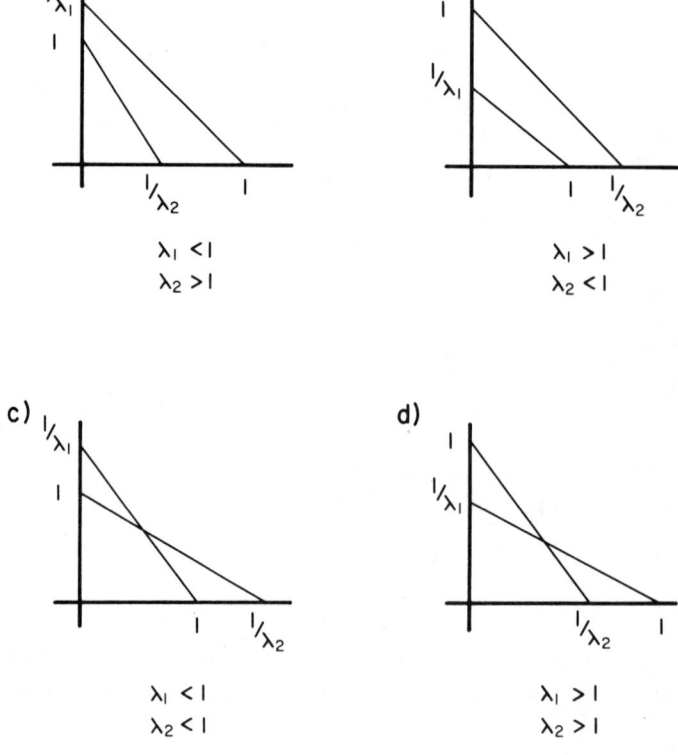

FIG. 2.1. *The possible isoclines for the system* (2.1).

In cases c and d, a critical point exists interior to the positive quadrant, so periodic solutions are not automatically ruled out. The Dulac criterion may be used to eliminate this possibility by choosing $\beta(x, y) = 1/xy$. Then

$$\frac{\partial}{\partial x}\left[\left(\frac{1}{xy}\right)x(1 - x - \lambda_1 y)\right] = -\frac{1}{y},$$

$$\frac{\partial}{\partial y}\left[\left(\frac{1}{xy}\right)ry(1 - y - \lambda_2 x)\right] = -\frac{r}{x}.$$

In the positive quadrant these quantities are negative so the Dulac criterion says that there are no periodic solutions there. If (x^*, y^*) is a solution of (2.2), then the variational matrix A has the form

$$\begin{bmatrix} -x^* & -\lambda_1 x^* \\ -r\lambda_2 y^* & -ry^* \end{bmatrix}.$$

If $x^* > 0$ and $y^* > 0$ then the trace of A is negative, so the sum of the eigenvalues is negative. The determinant is $rx^*y^*(1 - \lambda_1\lambda_2)$. Hence E_3 is an asymptotically stable attractor if $\lambda_1\lambda_2 < 1$ and a saddle point if $\lambda_1\lambda_2 > 1$. (Note that if $\lambda_1\lambda_2 = 1$ the lines in (2.2) are parallel or coincident. If coincident, there is a line of critical points; if parallel, the arguments given for cases a and b apply with stability replacing asymptotic stability in the argument. This zero eigenvalue case is considered "unlikely" and therefore "uninteresting.")

In case c, $1 - \lambda_1 > 0$ so E_1 is a saddle point, and $1 - \lambda_2 > 0$ so E_2 is a saddle point. Since $\lambda_1\lambda_2 < 1$, E_3 is an attractor. Therefore, all solutions with initial conditions in the open first quadrant tend to E_3. In this case the two competitors coexist.

In case d, E_1 and E_2 are asymptotically stable attractors and E_3 is a saddle point. Only two trajectories (the stable manifold of E_3) may have E_3 in their omega limit sets, so every other trajectory with initial conditions in the positive quadrant tends to one of the critical points on the boundary. Whether the limit is E_1 or E_2 depends on the initial conditions. Competitive exclusion holds, but the initial conditions determine the winner. Figure 2.2 shows a sketch of the phase plane for the four cases.

The next question of interest is to see what the above analysis has to do with the biology. There are many experimental attempts to verify the mathematics. We reproduce two figures from Hutchinson [40] of experiments of Gause [25]. Figure 2.3 shows the competition of Paramecium aurelia with Glaucoma scintillens. Figure 2.3a shows the time course of P. aurelia alone and with Glaucoma; 2.3b shows Glaucoma alone and with P. aurelia; 2.3c shows a typical trajectory in a phase plane plot. This is an example of case b above. Figure 2.4 shows the coexistence obtained in case c above with Paramecium bursaria and Paramecium aurelia. The top figure gives the time course and the bottom a phase plane plot of some trajectories in the experiment. There are some difficulties with this experiment, however. P. bursaria could enter a zone at the bottom of the culture where P. aurelia could not survive. Thus the competition was not perfect. It is not unusual that biological experiments which show coexistence of the competitors are criticized as "imperfect" competi-

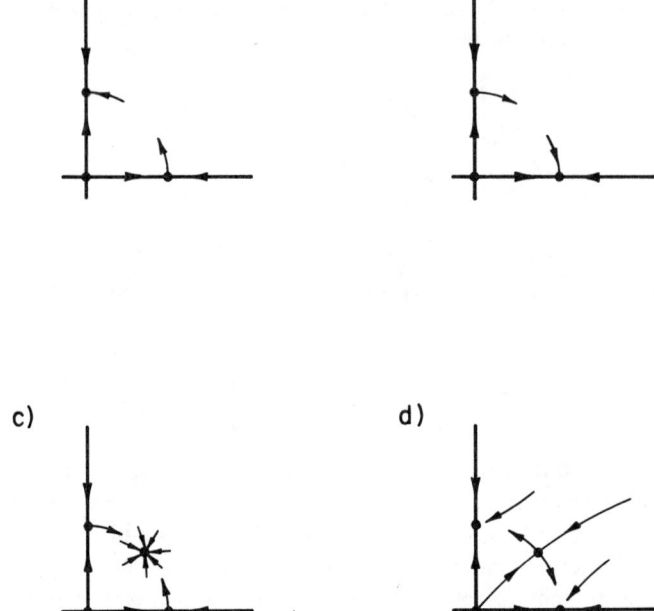

FIG. 2.2. *Phase plane sketch for the system* (2.1) *showing the four possible cases.*

tion. Whether "competitive exclusion" holds in nature is a matter of "interpretation." For an interesting discussion of this, see Hutchinson [40, p. 198].

Models with more general nonlinearities than those in (2.1) have been studied. For example, the competitive aspects may be modeled by

$$x' = xf(x, y), \qquad y' = yg(x, y)$$

where f is decreasing in y and g is decreasing in x. See, for example, Ayala, Gilpin and Ehrenfeld [4], Albrecht, Gatzke, Haddad and Wax [1], Freedman [23, Chap. 8], Hirsch and Smale [32, Chap. 12] for an introduction to the literature. We note only one of the many difficulties that are introduced—that there may now be many interior critical points. However, a general convergence property is preserved by the above monotonicity condition; see Hirsch [31].

3. Competition in the chemostat. The chemostat was described in considerable detail in Chapter 1. Here we consider exactly the same laboratory set-up as discussed there, except that two types of microorganisms are placed in the culture vessel instead of one. It is assumed that the two organisms do not affect each other in any way. For example, no toxins are produced and the metabolic products of one do not affect the other (or the chemostat is operated at a turnover rate sufficiently high that metabolic products wash out before they are significant). The two organisms compete only by consuming the common resource. This type of competition is called *exploitative* in contrast to interference competition which might occur if one

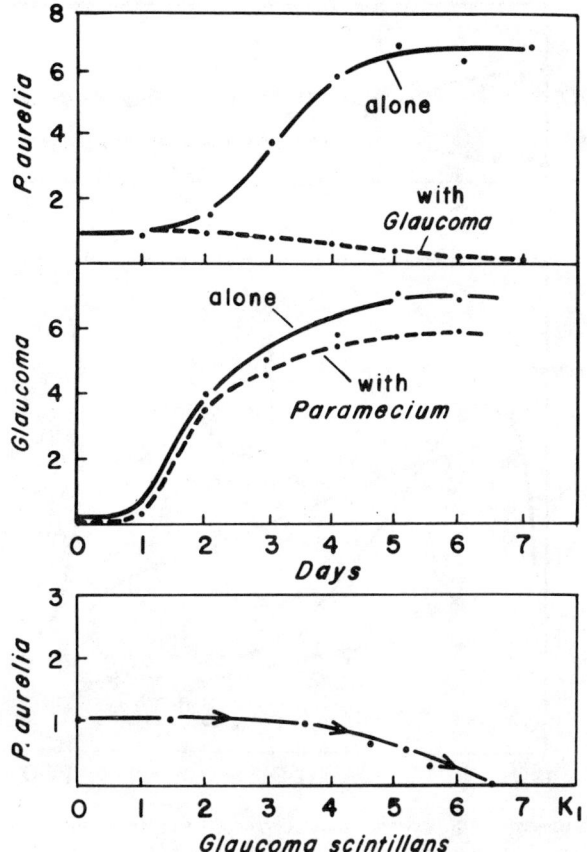

FIG. 2.3. *Upper panels, time course of cultures of P. aurelia and Glaucoma scintillans. Lower, phase plane plot from data, showing exclusion. From Hutchinson* [40], *using data of Gause.* [Copyright 1978, Yale University Press. Used by permission.]

organism produced a toxin against the other. The same type of normalization that produced (2.1) of Chapter 1 yields the system

(3.1)
$$S' = 1 - S - \frac{m_1 x S}{a_1 + S} - \frac{m_2 y S}{a_2 + S},$$

$$x' = \frac{m_1 x S}{a_1 + S} - x, \quad y' = \frac{m_2 y S}{a_2 + S} - y, \quad S(0) \geq 0, \quad x(0) > 0, \quad y(0) > 0.$$

This model differs from that considered in the previous section in that the resource being competed for is a variable in the system. On the other hand, it is very specific for this situation, while the Lotka–Volterra model is quite general.

It may happen that a competitor would not be able to survive in the chemostat, even if it were the only organism in the culture vessel. The omega limit set of a trajectory of the system (3.1) would then be trajectories of the simple chemostat

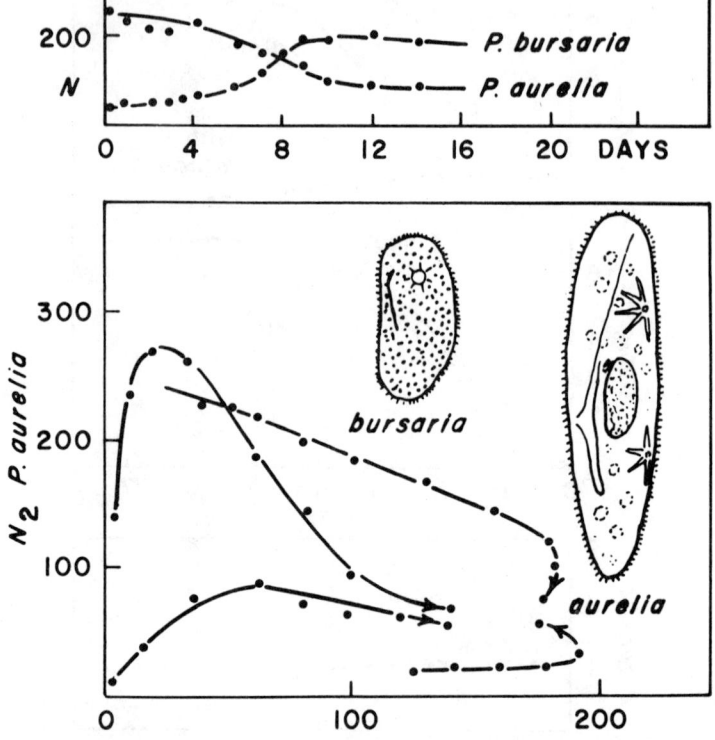

FIG. 2.4. *Upper, time course of P. aurelia and P. bursaria in a common culture, showing coexistence. Lower, phase plane plot from data. From Hutchinson [40], using data of Gause.* [Copyright 1978, Yale University Press. Used by permission.]

discussed in Chapter 1. This situation is covered in the next theorem. Define, for $m_i > 1$,

$$\lambda_i = \frac{a_i}{m_i - 1}$$

as in Chapter 1.

THEOREM 3.1 (inadequate competitors). *If $m_1 \leq 1$ then $\lim_{t \to \infty} x(t) = 0$. If $m_1 > 1$ and $\lambda_1 \geq 1$, then $\lim_{t \to \infty} x(t) = 0$. Similar statements hold for m_2, λ_2, and y.*

It is reasonable then to assume

(H) $\qquad m_i > 1, \quad i = 1, 2, \quad 0 < \lambda_1 < \lambda_2 < 1.$

The statement $\lambda_1 < \lambda_2$ is merely a choice of labeling.

THEOREM 3.2 (competitive exclusion). *Let* (H) *hold. Then*

$$\lim_{t \to \infty} S(t) = \lambda_1, \quad \lim_{t \to \infty} x(t) = 1 - \lambda_1, \quad \lim_{t \to \infty} y(t) = 0$$

for every solution of (3.1).

An early reference to the statement of competitive exclusion in the chemostat is Powell [53]. More recent work is Stewart and Levin [59], Hsu, Hubbell and Waltman [37] and Hsu [35]. Before giving a proof we simplify the problem somewhat. Note first that the positive octant is positively invariant. Moreover if $\Sigma(t) = 1 - S(t) - x(t) - y(t)$, then the system (3.1) can be written

$$\Sigma' = -\Sigma,$$

(3.1)'
$$x' = x\left[\frac{m_1(1 - x - y - \Sigma)}{1 + a_1 - x - y - \Sigma} - 1\right],$$

$$y' = y\left[\frac{m_2(1 - x - y - \Sigma)}{1 + a_2 - x - y - \Sigma} - 1\right], \quad \Sigma(0) > 0, \quad x(0) > 0, \quad y(0) > 0.$$

Clearly, $\Sigma = \Sigma(0) \exp(-t)$. This says that the omega limit set of any trajectory with initial conditions in the positive octant lies in the plane $\Sigma = 0$. Since every such trajectory will be asymptotic to its omega limit set, we analyze (3.1)' with $\Sigma = 0$. By arguments like those in §3 of Chapter 1, $(0, 0, 0)$ cannot be an omega limit point of a trajectory of (3.1)'. (Actually, the monotonicity is irrelevant, but it makes for an easy argument.) Using $\Sigma = 0$, we analyze

(3.2)
$$x' = x\left[\frac{m_1(1 - x - y)}{1 + a_1 - x - y} - 1\right],$$

$$y' = y\left[\frac{m_2(1 - x - y)}{1 + a_2 - x - y} - 1\right], \quad x(0) > 0, \quad y(0) > 0.$$

Note that the denominators are positive and that all trajectories are bounded. The system (3.2) has three critical points which we label as

$$E_0: (0, 0), \quad E_1: (0, 1 - \lambda_2), \quad E_2: (1 - \lambda_1, 0).$$

The statement of the theorem is equivalent to showing that all solutions tend to E_2. Since none of the critical points is interior to the positive quadrant, there are no limit cycles. The lines $x + y = 1 - \lambda_1$ and $x + y = 1 - \lambda_2$ divide the plane into three regions, and the direction of the vector field in each is shown in Fig. 3.1. At E_0 the variational matrix takes the form

$$\begin{bmatrix} \dfrac{(m_1 - 1)(1 - \lambda_1)}{1 + a_1} & 0 \\ 0 & \dfrac{(m_2 - 1)(1 - \lambda_2)}{1 + a_2} \end{bmatrix}.$$

Both eigenvalues are positive so the origin is a repeller and cannot be in the omega limit set of any noncritical trajectory. At E_1 the variational matrix takes the form

$$\begin{bmatrix} \dfrac{(m_1 - 1)(\lambda_2 - \lambda_1)}{a_1 + \lambda_2} & 0 \\ \dfrac{(\lambda_2 - 1)m_2 a_2}{(a_2 + \lambda_2)^2} & \dfrac{(\lambda_2 - 1)m_2 a_2}{(a_2 + \lambda_2)^2} \end{bmatrix}.$$

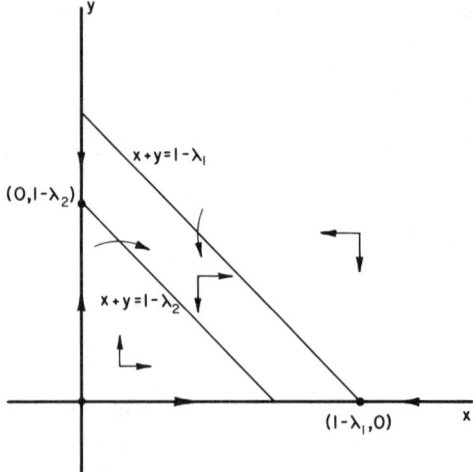

FIG. 3.1. *The vector field for the system* (3.2).

One eigenvalue is positive and one is negative, so the critical point is a saddle point. Moreover, the stable manifold lies on the y-axis and hence this critical point cannot be the omega limit set of any trajectory with initial conditions in the positive quadrant. The remaining critical point E_2 is (locally) asymptotically stable, as may be seen from the variational matrix

$$\begin{bmatrix} \dfrac{(\lambda_1 - 1)(m_1 a_1)}{(a_1 + \lambda_1)^2} & \dfrac{(\lambda_1 - 1)m_1 a_1}{(a_1 + \lambda_1)^2} \\ 0 & \dfrac{(m_2 - 1)(\lambda_1 - \lambda_2)}{a_1 + \lambda_2} \end{bmatrix}.$$

The Poincaré–Bendixson theorem then implies that all solutions of (3.2) have $(1 - \lambda_1, 0)$ as their omega limit set. Thus all trajectories of (3.1) tend to $(\lambda_1, 1 - \lambda_1, 0)$ as t tends to infinity. This proves Theorem 3.2.

If $\lambda_1 = \lambda_2$ then the two isoclines are coincident and there is a line of critical points. These critical points would correspond to the coexistence of the predators. This is the only case where coexistence is possible.

There remains to show how this result compares with the biology. Hansen and Hubbell [30] performed a series of experiments which compared predicted and actual outcomes. Before describing these experiments the predictive value of this theorem should be noted, particularly in contrast to the Lotka–Volterra model described in the previous section. In the Lotka–Volterra model there is a term involving the product xy. To measure the coefficient of this term, it would be necessary to grow the two organisms together. On the other hand, the terms in (3.1) involve only the competitor and the resource. It is possible to measure the coefficients by growing each organism separately on the nutrient. Once the coefficients are measured, Theorem 3.2 gives a prediction as to who will win the competition. Measuring the a_i's and the m_i's in the Michaelis–Menten dynamics is a standard procedure.

To understand the organization of the experiments, note that there are two obvious competing hypotheses for determining the winner of the competition. It might be suspected that the organism with the largest value of m should win since it reproduces better, or it might be suspected that the organism with the smallest a should win since it will reach more of its growth potential at a lower nutrient concentration. In the first experiment, the m_i's are approximately the same, but $a_1 < a_2$. In the second experiment $a_1 = a_2$ but $m_1 > m_2$. Finally (as described below) it was arranged that both the a's and the m's are different but $\lambda_1 = \lambda_2$.

The chart in Fig. 3.2 (reproduced from [30]) shows the organisms used, the run parameters (recall that $S^{(0)}$ and D are under the control of the experimenter), and

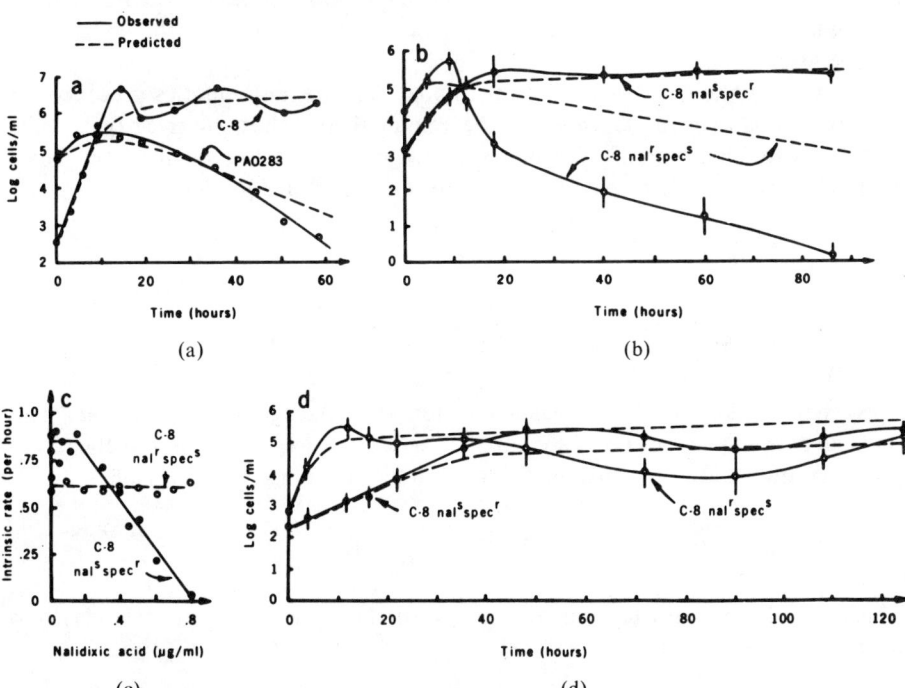

FIG 3.2. *Upper, parameters for the three experiments. Lower,* a *and* b *show the predicted (dashed lines) and experimental (solid lines) outcomes of chemostat experiments with competitive exclusion;* c *shows the effect of naladixic acid on the intrinsic rate of increase of the two competitors in the final experiment;* d *indicates coexistence when the proper amount of naladixic acid is added to the culture. From Hansen and Hubbell [30]. [Copyright 1980, American Association for the Advancement of Science. Used by permission.]*

the measured quantities. There are some differences in notation: $r = m - D$, $J = \lambda$, $K = a$, $\mu = m$, $S_0 = S^{(0)}$. The figures use unscaled variables. In Figs. 3.2 a, b and d, the dashed lines show the predicted outcome and the solid lines a curve fitted through the data points. Starting initial conditions were prejudiced in favor of the predicted loser. Figures 3.2a and b show that the predicted winner did indeed win the competition. Although this is quite a good fit for biological data, two discrepancies stand out. First of all, the theory predicts that the winner should approach the steady state monotonely, but the experiments showed damped oscillations. This may be due to delays which are in the system and not in the model. Secondly, the losing competitor always lost faster than predicted. This could be due to a dead zone near zero which is not reflected in the Michaelis–Menten dynamics, or perhaps to some necessary "critical mass" that must be present for an organism to grow. The important point is that the λ-parameter predicted the winner of the competition.

The organisms used in the second experiment had different sensitivity in their values of m to a chemical, naladixic acid. One was relatively insensitive to the presence of the chemical while the other's maximal growth rate dropped off markedly when the concentration increased. By adding the proper amount, it was possible to achieve equal values of λ even though the m's and a's were different. The sensitivity of the organisms to the chemical is shown in Fig. 3.2c and the result of the experiment shown in Fig. 3.2d. Coexistence is predicted and the experiment supports this.

For mathematicians it is of interest to note that the mathematics preceded the biology. This is an example where the theorem motivated the definitive experiment.

The above simple proof does not work for more than two competitors, but a similar theorem is true. Consider the system of differential equations

(3.3)
$$S' = 1 - S - \sum_i \frac{m_i x_i S}{a_i + S},$$
$$x_i' = x_i \left[\frac{m_i x_i S}{a_i + S} - D_i \right], \quad S(0) \geq 0, \quad x_i(0) > 0, \quad i = 1, 2, \cdots, n.$$

This system allows differing washout rates as well as more competitors. This washout rate can be viewed as a combination of the washout rate of the chemostat plus a natural death rate which may vary with the type of competitor. Define

$$\lambda_i = \frac{a_i D_i}{m_i - D_i}.$$

Theorem 3.1 continues to hold for (3.3) with this new definition of λ. (H) now becomes

(H)' $\qquad m_i > D_i, \quad 0 < \lambda_1 < \lambda_2 \leq \lambda_3 \leq \cdots \leq \lambda_n.$

THEOREM 3.3. *Let* (H)' *hold. Then*

$$\lim_{t \to \infty} S(t) = \lambda_1, \quad \lim_{t \to \infty} x_1(t) = 1 - \lambda_1, \quad \lim_{t \to \infty} x_i(t) = 0, \quad i = 2, 3, \cdots, n.$$

The proof uses a Lyapunov function [46] and may be found in Hsu [35].

The case where $S^{(0)}$ is a periodic function of time is also of interest, since this corresponds to the nutrient concentration fluctuating, say with the seasons, in a lake. This problem has been studied by Hsu [36], Smith [56], and Hale and Somolinas [29].

4. A three-level food chain. Consider the simple chemostat described in Chapter 1, consisting of a nutrient and an organism growing on that nutrient. Instead of adding a competing microorganism, as was done in the previous section, we add something which preys on the growing organism. The culture vessel then would contain a nutrient, an organism growing on that nutrient (which will be called the prey), and an organism which feeds on the prey (which will be called the predator). Such a system can be made of a sugar, a type of bacteria, and a type of ciliate. This is a simple example of a food chain. In nature a food chain would have many levels and many organisms on each level. Such a simple system as that described above is of interest because the mathematics will be tractable and experiments are possible in the laboratory.

It will be assumed that the functional response of the predator feeding on the prey will be of Michaelis–Menten type. For larger organisms this is also known as a Holling type II functional response [33]. Using the same nondimensionalization as before leads to a system of equations of the form

(4.1)
$$S' = 1 - S - \frac{m_1 x S}{a_1 + S},$$
$$x' = \frac{m_1 x S}{a_1 + S} - x - \frac{m_2 y x}{a_2 + x},$$
$$y' = \frac{m_2 x y}{a_2 + x} - y, \quad S(0) \geq 0, \quad x(0) > 0, \quad y(0) > 0.$$

Food chains have a large and extensive literature. The special case (4.1) has been studied by Canale [12], [13], Jost, Drake, Fredrickson and Tsuchiya [42], Sell [55], and Butler, Hsu and Waltman [9]. As before, an integral representation of (4.1) will show that the positive octant is positively invariant. Moreover, if $\Sigma(t) = 1 - S(t) - x(t) - y(t)$, then (4.1) may be rewritten

$$\Sigma' = -\Sigma,$$

(4.1)'
$$x' = x \left[\frac{m_1(1 - x - y - \Sigma)}{1 + a_1 - x - y - \Sigma} - 1 - \frac{m_2 y}{a_2 + x} \right],$$
$$y' = y \left[\frac{m_2 x}{a_2 + x} - 1 \right].$$

Solving the first equation in the system (4.1)' leads to the conclusion that the omega limit set of any trajectory of (4.1)' lies in the plane $\Sigma = 0$ or the omega limit set of any trajectory of (4.1) lies in the plane $S + x + y = 1$. We seek to analyze the behavior of the reduced system ($\Sigma = 0$). Before doing this we note that it is possible that the prey cannot live on the nutrient level or the predator cannot survive on the maximum attainable prey level. As before, let

$$\lambda_i = \frac{a_i}{m_i - 1}$$

when $m_i > 1$.

LEMMA 4.1. *If $m_1 \leq 1$ or if $\lambda_1 \geq 1$ ($m_1 > 1$) then $\lim_{t \to \infty} x(t) = 0$ (and necessarily $y(t)$ must also go to zero). If $m_2 \leq 1$ or if $\lambda_2 \geq 1$ ($m_2 > 1$) then $\lim_{t \to \infty} y(t) = 0$.*

The proof follows from differential inequalities. Because of the lemma we can assume

(H1) $\qquad\qquad m_i > 1 \quad \text{and} \quad 0 < \lambda_i < 1, \qquad i = 1, 2.$

Now, letting $\Sigma = 0$, ($S = 1 - x - y$), one obtains the equation of trajectories in the omega limit set as a two-dimensional system to which our techniques can be applied:

$$(4.2) \qquad x' = x\left[\frac{m_1(1 - x - y)}{1 + a_1 - x - y} - 1 - \frac{m_2 y}{a_2 + x}\right], \qquad y' = y\left[\frac{m_2 x}{a_2 + x} - 1\right].$$

There are three critical points which we label
E_0: $(0, 0)$,
E_1: $(1 - \lambda_1, 0)$,
E_2: (x_c, y_c), where (x_c, y_c) satisfies

$$(4.3) \qquad \frac{m_1(1 - x - y)}{1 + a_1 - x - y} - \frac{m_2 y}{a_2 + x} = 1, \qquad \frac{m_2 x}{a_2 + x} = 1$$

if there is such a solution in the positive quadrant.

The linearization of (4.1) (the variational matrix) is of the form

$$M = \begin{bmatrix} m_{11} & m_{12} \\ m_{21} & m_{22} \end{bmatrix},$$

where

$$m_{11} = \frac{m_1(1 - x - y)}{1 + a_1 - x - y} - \frac{m_2 y}{a_2 + x} + x\left[\frac{-m_1 a_1}{(1 + a_1 - x - y)^2} + \frac{m_2 y}{(a_2 + x)^2}\right],$$

$$m_{12} = \frac{-m_1 a_1 x}{(1 + a_1 - x - y)^2} - \frac{m_2 x}{a_2 + x}, \qquad m_{21} = \frac{m_2 a_2 y}{(a_2 + x)^2}, \qquad m_{22} = \frac{m_2 x}{a_2 + x} - 1.$$

At E_0 the matrix M takes the form

$$\begin{bmatrix} \dfrac{(m_1 - 1)(1 - \lambda_1)}{1 + a_1} & 0 \\ 0 & -1 \end{bmatrix}.$$

Thus under our hypothesis that $\lambda_1 < 1$, E_0 is a saddle point whose stable and unstable manifolds lie along the axes. The stable manifold represents the fact that the predator will die if there is no prey, and the unstable manifold, the fact that the prey will grow to $1 - \lambda_1$ if there are no predators.

At E_1 the matrix M takes the form

$$\begin{bmatrix} \dfrac{a_1 m_1(\lambda_1 - 1)}{(a_1 + \lambda_1)^2} & \dfrac{m_2(\lambda_1 - 1)}{1 + a_1 - \lambda_1} + \dfrac{a_1 m_1(\lambda_1 - 1)}{(a_1 + \lambda_1)^2} \\ 0 & \dfrac{(m_2 - 1)(1 - \lambda_1 - \lambda_2)}{1 + a_1 - \lambda_1} \end{bmatrix}$$

Thus E_1 is asymptotically stable if $\lambda_1 + \lambda_2 > 1$, and a saddle point if $\lambda_1 + \lambda_2 < 1$. An asymptotically stable E_1 corresponds to extinction of the predator at the top level of the food chain.

We turn now to solving (4.3). Since the second equation involves only x, a solution can be found easily—$x_c = \lambda_2$. Thus y_c must solve (rearranging terms)

$$(m_1 - 1)(1 - \lambda_1 - \lambda_2 - y_c)\lambda_2 = y_c(1 + a_1 - \lambda_2 - y_c).$$

If $\lambda_1 + \lambda_2 > 1$ there can be no positive solution since $x_c + y_c \leq 1$. Note that this is exactly the condition which makes E_1 asymptotically stable. If E_1 is a saddle point ($\lambda_1 + \lambda_2 < 1$) then the interior critical point must exist. Thus we will assume

(H2) $\qquad\qquad\qquad\qquad \lambda_1 + \lambda_2 < 1.$

At E_2 the matrix M takes the form

$$\begin{bmatrix} \dfrac{-m_1 a_1 \lambda_2}{(1 + a_1 - \lambda_1 - y_c)^2} + \dfrac{m_2 \lambda_2 y_c}{(a_2 + \lambda_2)^2} & \dfrac{-m_1 a_1 \lambda_2}{(1 + a_1 - \lambda_2 - y_c)^2} \\ \dfrac{y_c}{a_2 + \lambda_2} & 0 \end{bmatrix}.$$

Clearly det $M > 0$, so the eigenvalues have the same sign and, depending on the sign of the trace, E_2 is either a repeller or an asymptotically stable attractor. After a slight simplification the trace is

(4.4) $\qquad\qquad \dfrac{y_c}{m_2 \lambda_2} - \dfrac{m_1 a_1 \lambda_2}{(1 - a_1 - \lambda_2 - y_c)^2}.$

If (4.4) is positive, (x_c, y_c) is a repeller, and an application of the Poincaré–Bendixson theorem yields the existence of at least one limit cycle. If (4.4) is negative, then (x_c, y_c) is a local attractor. Two questions arise. If (x_c, y_c) is an attractor, is it a global attractor? If (x_c, y_c) is a repeller, is the resulting periodic solution unique? The first question is answered in the affirmative by the next theorem, but the second remains open.

THEOREM 4.1. *If* (H1) *and* (H2) *hold, then* (x_c, y_c) *exists in the first quadrant. If, in addition,* (4.4) *is negative, then* (x_c, y_c) *is a global attractor for the system* (4.2) *with initial conditions in the interior of the positive quadrant.*

The proof is long and tedious, involving several applications of Green's theorem; the reader is referred to [9] for details.

THEOREM 4.2. *Let the hypotheses of Theorem* 4.1 *hold. Then* $(1 - x_c - y_c, x_c, y_c)$ *is a global attractor for the system* (4.1).

Proof. We use (4.1)' instead of (4.1) so the critical point is of the form $(0, x_c, y_c)$. Let Ω denote the omega limit set of any trajectory Γ of (4.1)'. Ω lies in the plane $\Sigma = 0$. If the omega limit set contains a point in the open positive quadrant of the $\Sigma = 0$ plane, it must contain the point $(0, x_c, y_c)$. However, the linearization of (4.1)' about this point has eigenvalues with negative real part, namely -1 and the two eigenvalues associated with E_2 for (4.2). Hence it is an asymptotically stable local attractor and must be the only point of Ω. Therefore, we can assume that the Ω lies on the boundary of the first quadrant of $\Sigma = 0$. If Ω contains a point of the y-axis, it must contain the entire trajectory there which is unbounded. Hence any point of the y-axis is excluded. A similar contradiction follows if Ω contains a point of the x-axis to the right of E_1.

There remains only to show that E_0 and E_1 do not belong to Ω. While this could be done by an appeal to general theorems, we give a direct proof since it is straightforward. First of all E_0 is not the only point of Ω since Γ did not start on the stable manifold of Γ (i.e., $x(0) > 0$). Let B be a ball about the origin of radius r. Since the x-axis is the unstable manifold of E_0, one may cut a small closed neighborhood of the x-axis from the surface of the ball where all trajectories exit. If r is sufficiently small there is a sequence $t_n \to \infty$ as $n \to \infty$ such that $\Gamma(t_n)$ cuts the boundary of B from outside to inside. The sequence $\Gamma(t_n)$ lies in a compact set, is bounded away from the $y = 0$ plane, bounded away from the x-axis, and the component $\Sigma(t_n)$ tends to zero. Thus Ω has a point in the open quadrant of the plane $\Sigma = 0$. A similar argument can be made to exclude E_1. If Ω contains a point of the x-axis different

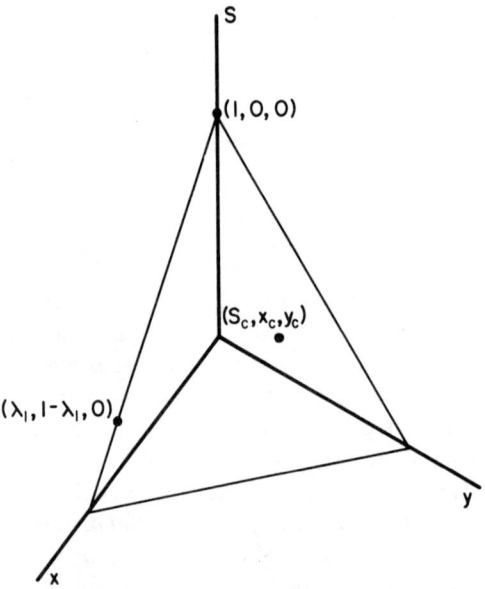

FIG. 4.1. *Location of the interior critical point for the system* (4.1).

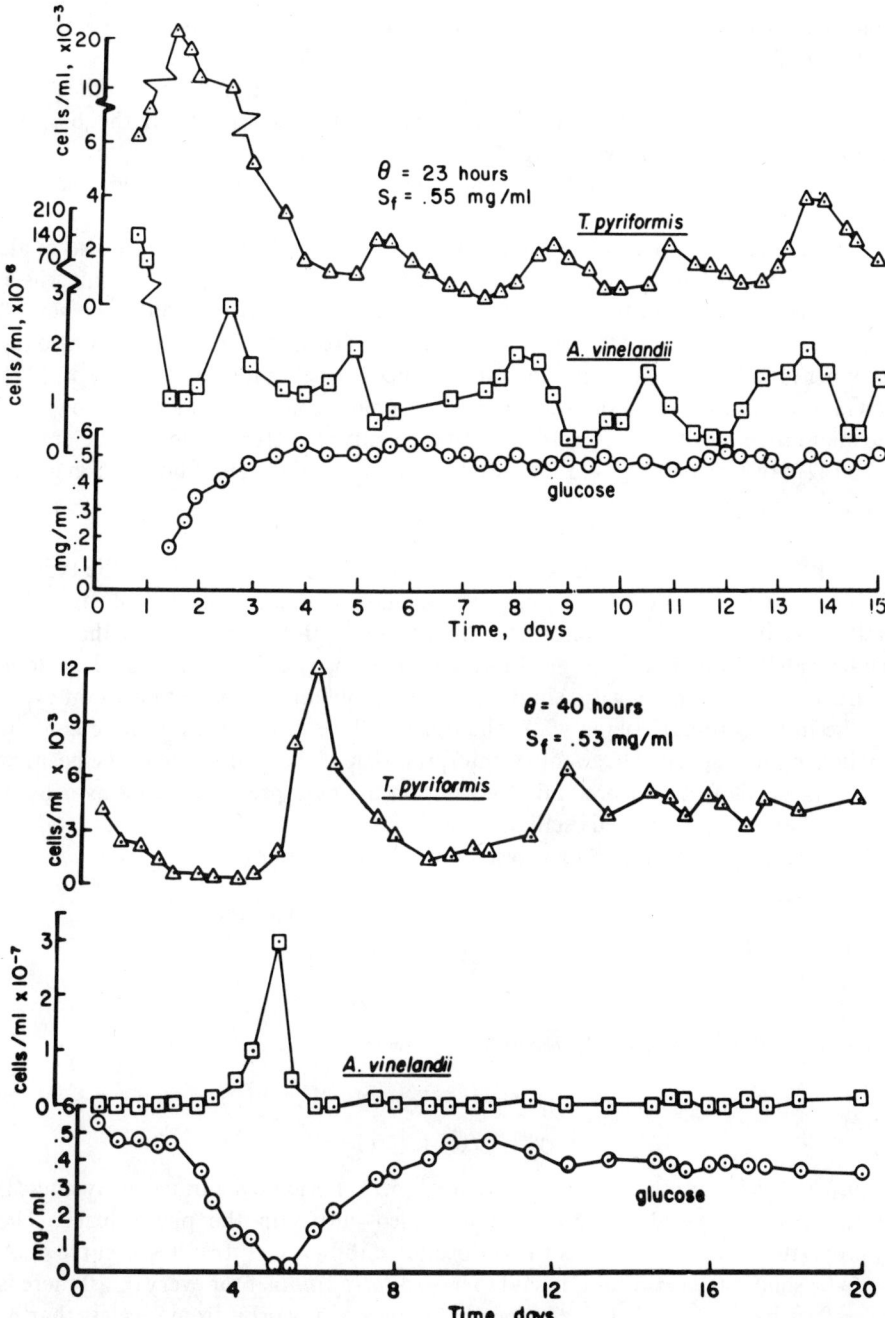

FIG 4.2. *The glucose-Azotobacter-Tetrahymena food chain. The upper panel shows sustained oscillations (indicative of a limit cycle) while the lower one shows damped oscillations (indicative of an asymptotically stable critical point). From Jost, Drake, Fredrickson and Tsuchiya [42]. [Copyright 1973, American Society for Microbiology. Used by permission.]*

from the two critical points it must also contain the critical points. This completes the proof.

In terms of the original system (4.1) there is a point (S_c, x_c, y_c), lying in the plane $S + x + y = 1$, which attracts all orbits with initial conditions in the positive octant. A schematic is shown in Fig. 4.1.

In case (4.2) has a unique limit cycle with a Floquet multiplier inside the unit interval, a similar theorem holds.

Both oscillatory and steady state cases seem to occur. In the paper of Jost et al. [42], when the chemostat was operated with one set of parameters, the limit was a steady state (achieved by damped oscillations) while, with a different set of parameters, sustained oscillations seemed to exist. (It should be noted that the data seem not to follow (4.1) exactly, in that the glucose level remained too high. In [42] this fact was used to substantiate the need for a more complicated model. However, the qualitative result is in agreement with the analysis presented here.) The graphs from [42] are shown in Fig. 4.2. The oscillations arise from a Hopf bifurcation in the plane $S + x + y = 1$.

5. Bifurcation from a limit cycle. We digress for one section from the type of problems being considered to discuss a new mathematical idea, the splitting of a limit cycle into two limit cycles as a parameter in the system passes through a critical value. Although this type of bifurcation is not as commonly studied as Hopf bifurcation, it is in many ways simpler. For the problems of interest one limit cycle will lie in a coordinate plane while the other will be in the open positive cone in Euclidean three-space. The eventual interpretation of the limit cycle in the positive cone will be that of two predators feeding on a single prey, surviving as a limit cycle—that is, as a sustained oscillation.

Consider the three-dimensional system, with a parameter v,

(5.1)
$$\begin{aligned} x' &= xf(x, y, z, v), \\ y' &= yg(x, y, z, v), \\ z' &= zh(x, y, z, v). \end{aligned}$$

If $(x(t), y(t))$ is a periodic solution of

(5.2)
$$\begin{aligned} x' &= xf(x, y, 0, v), \\ y' &= yg(x, y, 0, v). \end{aligned}$$

then $(x(t), y(t), 0)$ is a periodic solution of (5.1). Periodic solutions of dynamical systems, like those above, give rise to closed curves in the phase space. The appropriate stability concept is that of orbital stability. Call the closed curve C. A periodic solution generating C is said to be *orbitally stable* if for every $\varepsilon > 0$ there is a $\delta > 0$ such that if the distance in the phase space of any orbit from C is less than γ, then it remains within ε of C for all future time. If, in addition, the distance between C and the trajectory tends to zero as t tends to infinity, then the solution is said to be *asymptotically orbitally stable*. When the context is clear, "orbitally" may be dropped.

Suppose a periodic solution of (5.2) is asymptotically orbitally stable in the (x, y)-plane. Viewed as a solution of (5.1), the z direction near the orbit could be attracting or repelling as a function of the parameter v. Then, under appropriate hypotheses, it can happen that, as the stability in the z direction changes, the solution $(x(t), y(t), 0)$ splits into two closed orbits, one which remains in the (x, y)-plane and one which lies in the open positive octant of (x, y, z)-space. This splitting or bifurcation is the subject of this section. The analysis will depend on two ideas, a geometric one, the Poincaré map, and an analytic one, the Floquet exponents.

Let C be a limit cycle in the plane—a closed orbit. If W is a sufficiently small segment of a transversal to C at a point p, then one can define a mapping of a subset of W into W by associating to a point q of W the point $P(q)$ which is the first time the solution of the differential equation through q, say at time $t = 0$ cuts the transversal W. This is the first return map or the Poincaré map assoicated with the orbit and the point q. If the closed orbit is the trajectory of a three-dimensional system, the transversal above is replaced by a small planar section not containing the tangent vector. Although more complicated to visualize, the idea is the same—associate with a point of this planar set the first return of the orbit to the section. The important fact is that a fixed point of the Poincaré map corresponds to a periodic orbit of the original system. For example, if $(x(t), y(t), 0)$ is a periodic solution of (5.1), one can look for nearby periodic orbits by considering the fixed points of the Poincaré map associated with the given periodic orbit. If this fixed point should happen to have a nontrivial z component, then it must lie outside the (x, y)-plane.

The following, very complicated, theorem gives conditions for the existence of a curve of fixed points for a smooth mapping. In its use here the mapping will be the Poincaré map associated with a planar periodic orbit. The theorem except for the last statement is a special case of a theorem to be found in Marsden and McCracken [47].

THEOREM 5.1. *Let W be an open neighborhood of $0 \in R^2$ and let I be an open interval about $0 \in R$. Let P_v map W into R^2 such that $(v, x) \to P_v(x)$ is a C^r map for $r \geq 2$ from $I \times W$ into R^2 such that $P_v(0) = 0$, $v \in I$. Define $L_v = dP_v(0)$ and suppose that the eigenvalues of L_v lie inside the unit circle of the complex plane for $v < 0$. Assume that there is a real simple eigenvalue $k(v)$ of L_v such that $k(0) = 1$ and that $dk(0)/dv > 0$. Let u_0 be the eigenvector corresponding to $k(0)$. Then there is a C^{r-1} curve $\mathcal{C}$ of fixed points of P: $(v, x) \to (v, P_v)$ near $(0, 0)$ in $I \times R^2$ which together with $(v, 0)$ are the only fixed points of P near $(0, 0)$. The curve $\mathcal{C}$ is tangent to u_0 at $(0, 0)$.*

The difficulty in applying this theorem is that there is no way to obtain information about the linearization of the Poincaré map. There are theorems to tell us that the map exists for some region W and that it is differentiable (recall that we are assuming the right-hand sides of all our differential equations are continuously differentiable—usually they are real analytic). Help will come through the Floquet theory.

Consider a linear system of differential equations

(5.3) $$x' = A(t)x,$$

where A is an $n \times n$ matrix, $A(t + T) = A(t)$ and x is an n-dimensional vector. A major result in the theory of ordinary differential equations is the following:

THEOREM 5.2 (Floquet [16]). *A fundamental matrix $Y(t)$ for (5.3) can be represented in the form*

$$Y(t) = G(t) \exp(tL),$$

where L is a constant matrix and $G(t)$ is periodic with period T.

The eigenvalues of L are called the Floquet exponents and the eigenvalues of $\exp(TL)$ are called the Floquet multipliers. If the nonlinear equation

(5.4) $$x' = f(x)$$

has a periodic orbit $p(t)$, then the linearization of (5.4) about this orbit—called the variational equation—is of the form (5.3). The next theorem connects the two.

THEOREM 5.3 [16]. *Let $p(t)$ be a nonconstant periodic solution of period T of (5.4) and suppose that 0 is a simple Floquet exponent of the linearization about $p(t)$,*

$$x' = f_x(p(t))x,$$

and that all of the other exponents have negative real part. Then $p(t)$ is asymptotically orbitally stable.

The variational equation about a periodic solution $p(t)$ of (5.1) will be of the form (5.3) where $A(t)$ is a 3×3 matrix. Choose a point q on this orbit and let P be the Poincaré map in a planar cross section containing q and let dP be the linearization about q. The spectra of the two linearizations, the variational equation and the linearized Poincaré map, are connected by the following lemma.

LEMMA 5.4 [47]. *The spectrum of the linearization of the Poincaré map union $\{1\}$ is the set of Floquet multipliers of the variational equation.*

This lemma changes the question of the location of the eigenvalues of the linearization of the Poincaré map, needed to apply Theorem 5.1, to a question of determining the Floquet multipliers of the associated variational equation. In general these are equally intractable problems. However, if the system (5.1) is further specialized to the system

(5.5) $$x' = xf(x, y, z, v), \quad y' = yg(x, y, v), \quad z' = zh(x, z, v),$$

the problem becomes easier. The variational equation for (5.5) has the general form

$$\begin{bmatrix} f(x, y, z) + xf_x(x, y, z) & xf_y(x, y, z) & xf_z(x, y, z) \\ yg_x(x, y, z) & g(x, y) + yg_y(x, y) & 0 \\ zh_x(x, z) & 0 & h(x, z) + zh_z(x, z) \end{bmatrix}.$$

At a periodic solution of the form $(x(t), y(t), 0)$ this becomes

$$J(t) = \left[\begin{array}{cc|c} A(t) & & a(t) \\ & & 0 \\ \hline 0 & 0 & b(t) \end{array} \right].$$

There are now two linear systems of interest:

(5.6) $$\xi' = A(t)\,\xi$$

and

(5.7) $$\eta' = J(t)\,\eta.$$

Solutions of (5.7) may be represented by three linearly independent vectors. These can be chosen to be of the form

$$\begin{pmatrix} \eta_{11}(t) \\ \eta_{21}(t) \\ 0 \end{pmatrix} \begin{pmatrix} \eta_{12}(t) \\ \eta_{22}(t) \\ 0 \end{pmatrix} \begin{pmatrix} \eta_{13}(t) \\ \eta_{23}(t) \\ \eta_{33}(t) \end{pmatrix}.$$

Then since $\eta_{31}(0) = \eta_{32}(0) = 0$ implies that $\eta_{31}(t) \equiv \eta_{32}(t) \equiv 0$,

$$\begin{pmatrix} \eta_{11}(t) \\ \eta_{21}(t) \end{pmatrix} \begin{pmatrix} \eta_{12}(t) \\ \eta_{22}(t) \end{pmatrix}$$

solve (5.6). Thus

$$Y(t) = \begin{bmatrix} \eta_{11}(t) & \eta_{12}(t) \\ \eta_{21}(t) & \eta_{22}(t) \end{bmatrix}$$

is a fundamental matrix for (5.6) and

$$Z(t) = \left[\begin{array}{cc|c} Y(t) & & \eta_{31}(t) \\ & & \eta_{32}(t) \\ \hline 0 & 0 & \eta_{33}(t) \end{array}\right]$$

is a fundamental matrix for (5.6). The Floquet multipliers are the eigenvalues of the fundamental matrix, which is the identity at the origin, evaluated at the period. This follows readily from Theorem 5.2 since if $Y(t)$ is a fundamental matrix for (5.3) then

$$Y(T) = G(T)\exp(TL) = G(0)\exp(TL) = Y(0)\exp(TL)$$

and $Y(0)$ is the identity. Thus the Floquet multipliers of (5.6) and (5.7) are the eigenvalues of $Y(T)$ and $Z(T)$, respectively. Since

$$Z(T) = \left[\begin{array}{cc|c} Y(T) & & \eta_{31}(T) \\ & & \eta_{32}(T) \\ \hline 0 & 0 & \eta_{33}(T) \end{array}\right],$$

it follows that

$$\det[Y(T)] = \exp\left(\int_0^T \operatorname{tr} A(s)\,ds\right)$$

and

$$\det [Z(T)] = \exp\left(\int_0^T \operatorname{tr}[A(s) + b(s)]\, ds\right)$$

$$= \exp\left(\int_0^T \operatorname{tr} A(s)\, ds\right) \exp\left(\int_0^T b(s)\, ds\right).$$

Thus, the Floquet multipliers of (5.7) are the Floquet multipliers of (5.6) union $\exp(\int_0^T b(s)\, ds)$.

The basic idea in the next two sections is to vary the coefficients in the equation in such a way that the multiplier $\exp(\int_0^T b(s)\, ds)$, which will also be an eigenvalue of the linearization of the Poincaré map, will satisfy the conditions of Theorem 5.1. The multipliers associated with $Y(T)$ will be determined from a computation, or their behavior will be postulated.

6. Two predators competing for a renewable resource. We consider now a resource—hereafter referred to as the prey—which reproduces according to the logistic equation. We suppose that two predators feed exclusively upon this prey, the functional response being of Michaelis–Menten or (more appropriate in this context) Holling dynamics. The basic question is whether the coexistence of all three populations is possible. This is a fundamental question in ecology and there is considerable experimental evidence to suggest that competitive exclusion should hold for "perfect" competition. Sections 2 and 3 provided examples where competitive exclusion did hold. With a renewable resource as prey, the answer will be quite different. McGehee and Armstrong [50] have given a mathematical example showing that coexistence is possible, and numerical computations of Koch [45] and of Hsu, Hubbell and Waltman [38] have also supported coexistence. The material here follows [38] and Butler and Waltman [10]. It will also be assumed that the predators do not interfere with each other—they compete by each consuming the prey that would otherwise be available to its competitor. As noted before, this type of competition is referred to as exploitative.

The basic equations take the form

(6.1)
$$S' = rS\left(1 - \frac{S}{K}\right) - \frac{m_1 xS}{\gamma_1(a_1 + S)} - \frac{m_2 yS}{\gamma_2(a_2 + S)},$$

$$x' = \frac{m_1 xS}{a_1 + S} - D_1 x,$$

$$y' = \frac{m_2 yS}{a_2 + S} - D_2 y$$

with $S(0) > 0$, $x(0) > 0$ and $y(0) > 0$. Note that there is no conservation; S is growing on nutrients outside the system. The equations may be changed to nondimensional form as before, yielding

$$S' = S(1-S) - \frac{m_1 xS}{a_1 + S} - \frac{m_2 yS}{a_2 + S},$$

(6.2) $\quad x' = \dfrac{m_1 xS}{a_1 + S} - D_1 x,$

$\quad y' = \dfrac{m_2 yS}{a_2 + S} - D_2 y, \quad S(0) > 0, \quad x(0) > 0, \quad y(0) > 0.$

From the form of the equations it follows that the positive octant is invariant, and from some simple differential inequalities it follows that all solutions are bounded. The following lemma deals with the case of inadequate predators—predators which will not survive in the system even without competition.

THEOREM 6.1. *Let* $\lambda_i = a_i D_i/(m_i - D_i)$. *If* $m_1 \leq D_1$ *or if* $m_1 > D_1$ *and* $\lambda_1 \geq 1$, *then*

$$\lim_{t \to \infty} x(t) = 0.$$

If $m_2 \leq D_2$ *or if* $m_2 > D_2$ *and* $\lambda_2 \leq 1$ *then*

$$\lim_{t \to \infty} y(t) = 0.$$

The proof of Theorem 6.1 follows from simple arguments with inequalities. Just as in § 3 the consequence of this is that we may assume

(H1) $\qquad\qquad \lambda_i < 1, \quad m_i > D_i, \quad i = 1, 2, \qquad \lambda_1 < \lambda_2.$

LEMMA 6.2. *Let* (H1) *hold and let* $m_2/D_2 \leq m_1/D_1$. *Then*

$$\lim_{t \to \infty} y(t) = 0.$$

(*The convergence is exponential.*)

The proof of this lemma is very technical, and the reader is referred to [37], [39] for proof of a similar lemma. We note one special case, however. If $m_1 > m_2$, $D_1 < D_2$ and $a_1 < a_2$, then the x predator has an advantage over the y predator for all levels of the prey population, and so it is not surprising that the weak competitor is eliminated. The interesting case then is where the "advantage" shifts with the level of the prey. Essentially, the lemma applies when the change of advantage occurs at a sufficiently low level of the prey.

When the lemma does apply, the omega limit set of any trajectory of (6.2) is two-dimensional, lying in the (S,x)-plane. The asymptotic behavior of solutions then is determined by the behavior of the two-dimensional system

(6.3) $\qquad S' = S(1-S) - \dfrac{mxS}{a+S}, \qquad x' = \dfrac{mxS}{a+S} - Dx.$

The system (6.3) is a "typical" predator–prey system, and we proceed to analyze the asymptotic behavior of its trajectories. First of all, there are three critical points which we label as

$$E_0: (0,0), \quad E_1: (1,0), \quad E_2: \left(\lambda, \frac{\lambda(1-\lambda)}{D}\right)$$

where $x = aD/(m-D)$. The variational matrix (the coefficient matrix in the variational equation) takes the form

$$\begin{bmatrix} 1 - 2S - \dfrac{mxa}{(a+S)^2} & \dfrac{-mS}{a+S} \\[2mm] \dfrac{max}{(a+S)^2} & \dfrac{mS}{a+S} - D \end{bmatrix}.$$

At E_0 this is

$$\begin{bmatrix} 1 & 0 \\ 0 & -D \end{bmatrix}$$

so the origin is a saddle point with the stable and unstable manifolds lying along the coordinate axes. At E_1 the variational matrix takes the form

$$\begin{bmatrix} -1 & \dfrac{-m}{1+a} \\[2mm] 0 & \dfrac{(m-D)(1-\lambda)}{a+1} \end{bmatrix}.$$

Under hypothesis (H1) this too is a saddle point with the stable manifold lying along the x-axis. The biological interpretation of these two critical points is that, in the absence of a predator, the prey grows to its carrying capacity while, in the absence of the prey, the predator becomes extinct.

At E_2 the variational matrix takes the form

$$\begin{bmatrix} -\lambda\left[1 - \dfrac{1-\lambda}{a+\lambda}\right] & \dfrac{-m\lambda}{a+\lambda} \\[2mm] \dfrac{(m-D)(1-\lambda)}{a+\lambda} & 0 \end{bmatrix}.$$

The product of the eigenvalues of this matrix is positive while the sum of the eigenvalues may be written as $(1 - a - 2\lambda)D/m$. Thus if $1 - a - 2\lambda < 0$, the critical point is (locally) asymptotically stable, while if $1 - a - 2\lambda > 0$, the critical point is a repeller. (At $1 - a - 2\lambda = 0$ a Hopf bifurcation occurs.) In the original variables of (6.1) this would have taken the form $K > a + 2\lambda$ or $K < a + 2\lambda$. To have a global result for trajectories with initial conditions in the open positive

quadrant, one must show
  i) If $a + 2\lambda > 1$ there are no limit cycles.
  ii) If $a + 2\lambda < 1$ there is a unique limit cycle.
In case i) the only possible omega limit set is the critical point, while in case ii) the Poincaré–Bendixson theorem would say that the unique periodic trajectory is the omega limit set of all noncritical orbits.

LEMMA 6.3. *If $a + 2\lambda > 1$ and $\lambda < 1$, then the system (6.3) has no limit cycles in the open positive quadrant.*

Proof. The lemma follows from an application of the Dulac criterion using the function
$$\beta(S, x) = \left(\frac{S}{a+S}\right)^\alpha x^\delta$$
in the region $x > 0$ and $1 + 1/a > S > 0$ where α and δ are to be chosen (wisely)—see [39].

As noted above a consequence of this lemma and the Poincaré–Bendixson theorem is that the critical point E_2 is a global attractor of orbits of (6.3) with initial conditions in the open positive quadrant. When the inequality is reversed, E_2 is an unstable critical point and the Poincaré–Bendixson theorem and the boundedness of solutions yield the existence of at least one limit cycle.

LEMMA 6.4. *If $a + 2\lambda < 1$ there is a unique limit cycle for the system (6.3) and this limit cycle has a Floquet multiplier inside the unit circle.*

The proof of this lemma, which is quite difficult, may be found in Cheng [14]. It uses a clever symmetry argument.

THEOREM 6.5. *Let (H1) hold with $m_1/D_1 \geq m_2/D_2$. Then:*
1) *If $a_1 + 2\lambda_1 > 1$, every solution of (6.2) satisfies*
$$\lim_{t \to \infty} S(t) = \lambda_1, \quad \lim_{t \to \infty} x(t) = \frac{(1-\lambda_1)\lambda_1}{D_1}, \quad \lim_{t \to \infty} y(t) = 0.$$

2) *If $a_1 + 2\lambda_1 < 1$, every solution of (6.2) except one tends to the unique periodic orbit of the system (6.3).*

A proof appears in [39]. The exceptional orbit is the stable manifold of the critical point $(\lambda_1, (1-\lambda_1)\lambda_1/D_1, 0)$ of the system (6.2). The interesting remaining case is that $\lambda_1 < \lambda_2$, $a_1 < a_2$ and $m_1/D_1 < m_2/D_2$. (These are not independent conditions.)

The system (6.1) was simulated with a variety of coefficients to see if coexistence was possible. When all of the parameters are held fixed except a_1 and K, a diagram like Fig. 6.1 results. (This figure appeared in [38].) The unshaded region represents the region in parameter space where coexistence occurs. (The various lines in the figure refer to theorems established in [38] and [39].) Note that the coexistence regions come in a variety of shapes, sometimes robust and sometimes quite thin. The most interesting feature, however, was that the competitive outcome reversed—the λ-criterion alone did not determine the competitive outcome. Figure 6.2 [38], which is Fig. 6.1G with a_1 fixed, illustrates this point. As K is increased from zero, at first

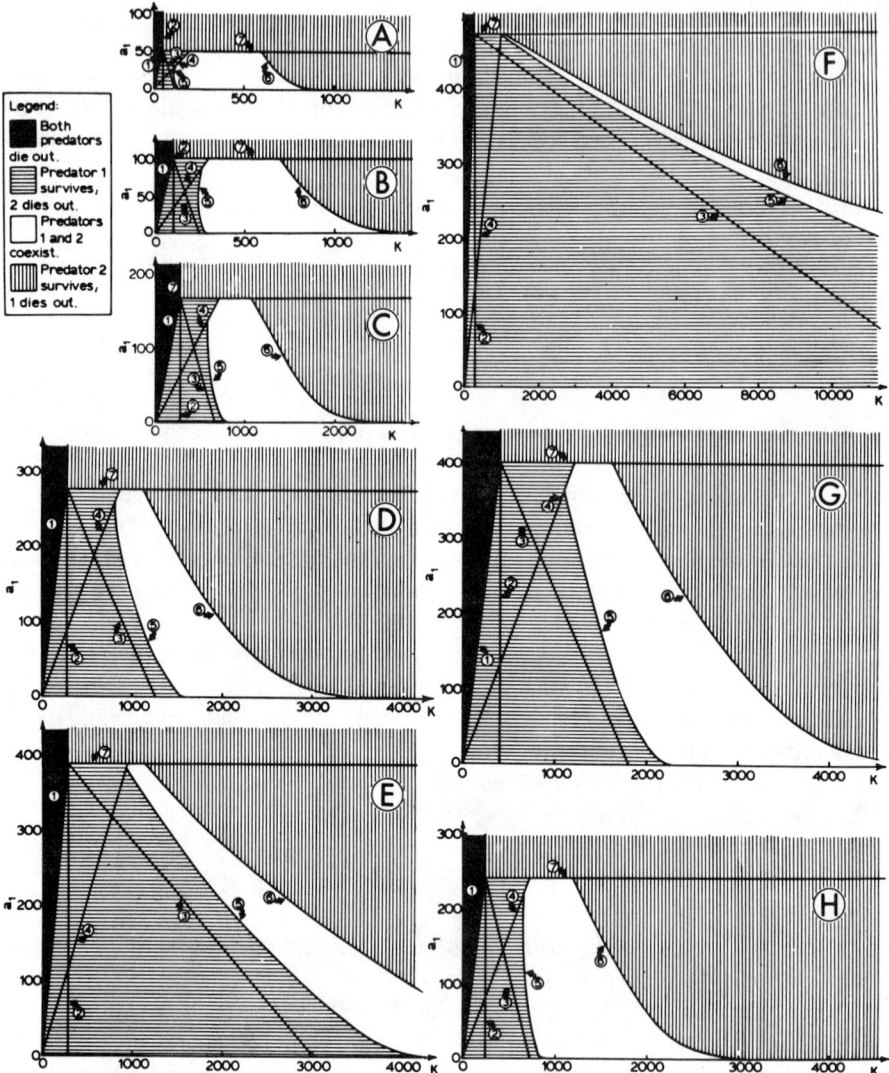

FIG. 6.1. *Simulation of* (6.1) *when all of the parameters except a_1 and K are held fixed. The unshaded region areas show the regions of coexistence for a variety of parameter values. From Hsu, Hubbell and Waltman* [38]. [Copyright 1978, Ecological Society of America. Used by permission.]

neither population survives, then only x survives, then x and y coexist, and finally only y survives. Note also that coexistence is not possible as a steady state—there is no critical point in the interior of the positive octant. Figures 6.3 and 6.4 show the actual orbits for the schematic in Fig. 6.2. They show quite dramatically that coexistence occurs as a *limit cycle*. These limit cycles were found by starting the simulation with arbitrary initial conditions and waiting for the trajectory to "settle down"—hence the cycles exhibited quite strong stability properties. Only one limit

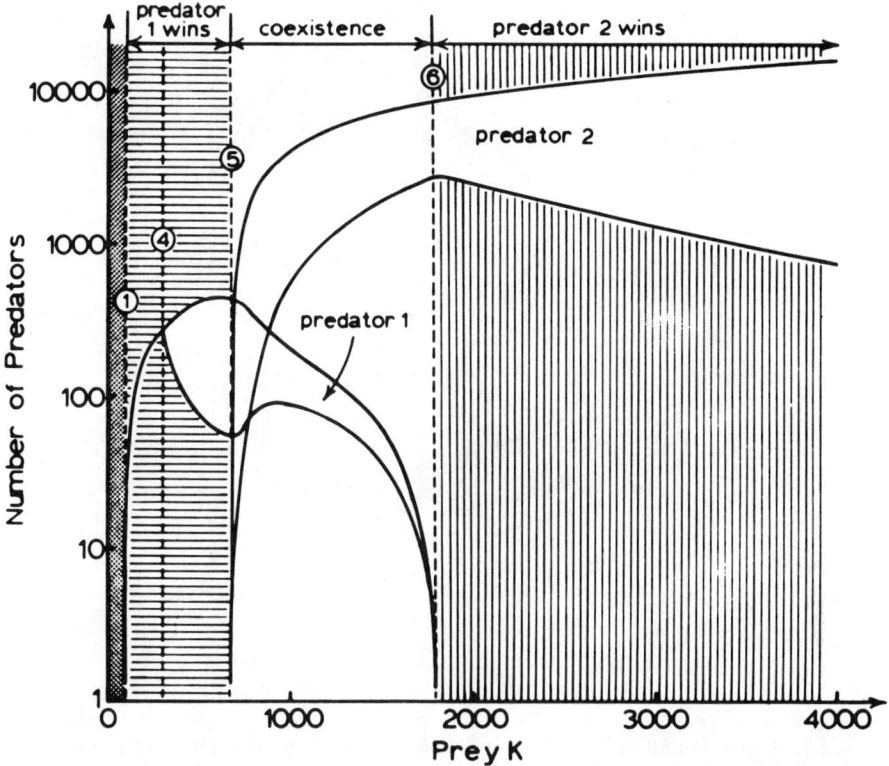

FIG. 6.2. *A cut through Fig. 6.1G with* $a_1 = 100$. *Other parameters are* $r = 20 \ln 2$, $D_1 = \ln 2/2$, $D_2 = \ln 2$, $\gamma_1 = .1$, $\gamma_2 = 1.14$, $a_2 = 720.$, $m_1 = \ln 2$, $m_2 = 4 \ln 2$. *The upper and lower curves measure the maximum and minimum of the oscillations. The middle area shows the coexistence region. From Hsu, Hubbell, and Waltman* [38]. [Copyright 1978, Ecological Society of America. Used by permission.]

cycle was found in the positive cone for each set of parameter values. (The author wishes to acknowledge with thanks the assistance of Dietmar Saupe who prepared Figs. 6.3 and 6.4.)

We now attempt to make the initial part of the above description rigorous, by showing the bifurcation into the coexistence region of parameter space. The simulation was performed with K as the bifurcation parameter. This is, conceptually, the correct choice since one can think of the organisms being fixed but the carrying capacity of the system being increased, perhaps by raising the underlying nutrient level for the prey population. However, for mathematical reasons, K is a poor choice. The methods of the previous section will be applied using the unique periodic orbit of (6.3) as the origin for the Poincaré map. However, if K is left in the system, as in (6.1), this orbit changes as K changes, making estimates difficult. On the other hand, if a_2 or D_2 is used as a bifurcation parameter, and K is scaled out of the system, as in (6.2), the periodic orbit of (6.3) does not change as the parameter is varied.

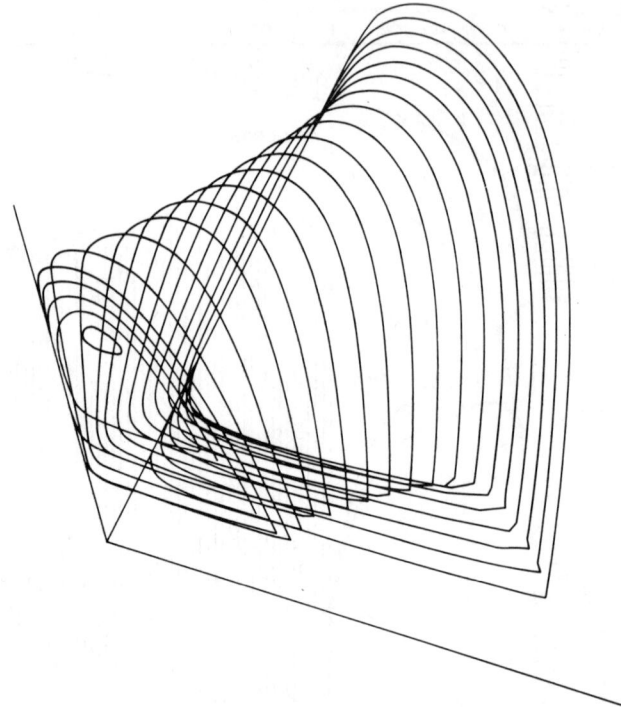

FIG. 6.3. *A three-dimensional plot showing that the limit cycle traverses the positive octant as K increases, collapsing into the (S, y)-plane after bifurcating from the (S, x)-plane. The parameters are as given in Fig. 6.2.*

THEOREM 6.6. *Let m_1, a_1, D_1 be fixed so that $m_1 > D_1$ and that $\lambda_1 < 1$ and $a_1 + 2\lambda_1 < 1$. Fix m_2 and D_2, $m_2 > D_2$, and $m_1/D_1 < m_2/D_2$. There exists a number $A > a_1$ such that for $a_2 < A$, $A - a_2$ small, it follows that $\lambda_1 \leq \lambda_2$ and that (6.2) has a periodic solution in the positive octant arbitrarily near the (S, x)-plane.*

Proof. Let $(S(t), x(t))$ be the unique periodic solution of (6.3). Let T denote the period of this solution and let ν be the Floquet multiplier with $|\nu| < 1$. $(S(t), x(t), 0)$ is a solution of (6.2). Linearize about this solution. The Floquet multipliers are $1, \nu$, $\mu = \exp(\int_0^T [m_2 S(t)/(a_2 + S(t))]dt - D_2 T)$. The linearization of the Poincaré map has eigenvalues ν and μ. Let

$$r(a) = \frac{m_2}{T} \int_0^T \frac{S(t)\, dt}{a + S(t)},$$

so

$$\mu = \exp(T(r(a_2) - D_2)).$$

Note that $S(t)$ is independent of a_2 and D_2. Also note that $r(0) = m_2 > D_2$ and that for a large, $r(a)$ is small. Hence there exists a number A such that $r(A) = D_2$. Furthermore

$$\frac{dr}{da} = -\frac{m_2}{T} \int_0^T \frac{S(t)\, dt}{(a + S(t))^2} < 0.$$

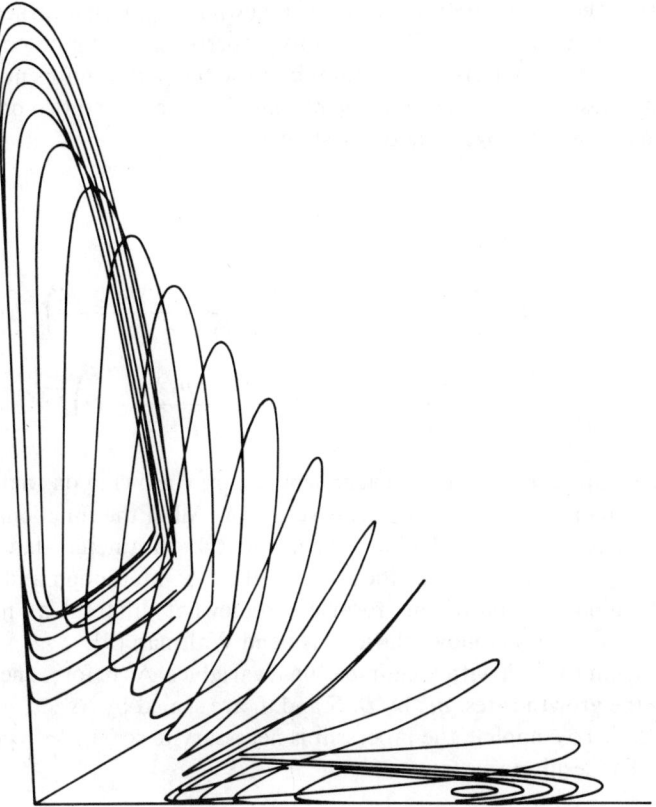

FIG. 6.4. *Another perspective of Fig. 6.3.*

Thus μ crosses the unit circle transversally at $a_2 = A$. Theorem 5.1 may now be applied to yield a curve of fixed points of the Poincaré map. These fixed points correspond to the periodic solutions claimed in the theorem. The eigenvector corresponding to the eigenvalue crossing the unit circle cannot lie in the (S, x)-plane, so the bifurcation is into the positive octant. (Note that there are also fixed points on the curve given by Theorem 5.1 corresponding to negative values of y, but these are biologically meaningless.)

The result is, of course, a local one—it has not been shown that it continues in the nice fashion of Fig. 6.3—nor has the stability been established. Under the additional hypothesis that $\lambda_2 - \lambda_1$ is small, these questions have been answered by Keener [43] and Smith [57] using asymptotic methods. Still, many interesting questions remain.

7. Two predators feeding on a prey growing in a chemostat. A single population growing on a constantly input nutrient in a chemostat was described in §3 of Chapter 1. Suppose now two predators feed exclusively on this population. Is coexistence possible? Note that this problem differs from that described in §3 of this chapter in that the competition is for a self-renewing population rather than for a nutrient input at a constant rate. It shares this property with the problem considered in §6 but differs from that problem in that, because the prey is nutrient limited, the system is now closed—all of the components, nutrient, prey, and two predators, are modeled. The equations governing this system are

$$S' = (S^{(0)} - S)D - \frac{m_1 S x}{\gamma_1(a_1 + S)},$$

(7.1)
$$x' = x\left(\frac{m_1 S}{a_1 + S} - D - \frac{m_2 y}{\gamma_2(a_2 + x)} - \frac{m_3 z}{\gamma_3(a_3 + x)}\right),$$

$$y' = y\left(\frac{m_2 x}{a_2 + x} - D\right), \quad z' = z\left(\frac{m_3 x}{a_3 + x} - D\right),$$

where, as in Chapter 1, S is the nutrient, x is the prey growing on that nutrient, y and z are the predators which feed exclusively on x. All of the functional responses are of Michaelis–Menten or Holling type, and the parameters have the same meaning as in §§3 and 4. The justification for the S-x interaction and for the x-y and x-z interactions to be of this form are somewhat different as noted in the introduction. This section follows Butler, Hsu and Waltman [9].

It is convenient to work with nondimensional variables. As before, one scales time by $1/D$ and the growth rates, m_i, by D. S and a_1 are scaled by $S^{(0)}$. x, a_2 and a_3 are scaled by $\gamma_1 S^{(0)}$. To complete the process it is necessary to scale y by $\gamma_1 \gamma_2 S^{(0)}$ and z by $\gamma_1 \gamma_3 S^{(0)}$. The resulting system is

$$S' = 1 - S - \frac{m_1 S x}{a_1 + S},$$

$$x' = x\left(\frac{m_1 S}{a_1 + S} - 1 - \frac{m_2 y}{a_2 + x} - \frac{m_3 z}{a_3 + x}\right),$$

(7.2)
$$y' = y\left(\frac{m_2 x}{a_2 + x} - 1\right),$$

$$z' = z\left(\frac{m_3 x}{a_3 + x} - 1\right),$$

It is this system of equations which will be analyzed. Note that all solutions which begin in the positive cone remain there for all future time.

The following lemma provides boundedness of solutions and reduces the problem from four dimensions to three.

LEMMA 7.1. *The omega limit set of any trajectory of* (7.2) *lies on the hyperplane* $S + x + y + z = 1$.

Proof. Let $\Sigma = 1 - S(t) - x(t) - y(t) - z(t)$. Then Σ satisfies $\Sigma' = -\Sigma$ and the lemma follows.

(7.2) can be written

(7.2)'
$$\Sigma' = -\Sigma,$$
$$x' = x\left(\frac{m_1(1-x-y-z-\Sigma)}{a_1+1-x-y-z-\Sigma} - 1 - \frac{m_2 y}{a_2+x} - \frac{m_3 z}{a_3+x}\right),$$
$$y' = y\left(\frac{m_2 x}{a_2 x} - 1\right),$$
$$z' = z\left(\frac{m_3 x}{a_3+x} - 1\right), \quad \Sigma(0) > 0, \quad x(0) > 0, \quad y(0) > 0, \quad z(0) > 0.$$

It is sufficient then to analyze solutions on the omega limit set (provided, of course, something turns out to be stable there). On this hyperplane trajectories satisfy

(7.3)
$$x' = x\left(\frac{m_1(1-x-y-z)}{1+a_1-x-y-z} - 1 - \frac{m_2 y}{a_2+x} - \frac{m_3 z}{a_3+x}\right),$$
$$y' = y\left(\frac{m_2 x}{a_2+x} - 1\right),$$
$$z' = z\left(\frac{m_3 x}{a_3+x} - 1\right), \quad x(0) > 0, \quad y(0) > 0, \quad z(0) > 0.$$

It can happen that the prey, x, cannot survive on the nutrient level being input, or that one of the predators, y or z, cannot survive on the prey level, independent of competition. These inadequacies are cared for in the following lemma.

LEMMA 7.2. *Let $\lambda_i = a_i/(m_i - 1)$. If $m_1 \leq 1$ or if $m_1 > 1$ and $\lambda_1 \geq 1$, then $\lim_{t \to \infty} x(t) = 0$ (and necessarily $y(t)$ and $z(t)$ tend to zero also). If $m_2 \geq 1$ or if $m_2 > 1$ and $\lambda_2 \geq 1$, $\lim_{t \to \infty} y(t) = 0$. If $m_3 \geq 1$ or if $m_3 > 1$ and $\lambda_3 \geq 1$, then $\lim_{t \to \infty} z(t) = 0$.*

If one of the predators becomes extinct (i.e., $y(t)$ or $z(t)$ tends to zero as t tends to infinity) then the resulting dynamical system is of the form considered in §4 and the asymptotic behavior is determined (see the summary below). Thus one may suppose

(H1) $\qquad\qquad m_i > 1, \quad \lambda_i < 1, \quad i = 1, 2, 3, \qquad \lambda_2 < \lambda_3.$

The last statement is just a choice of labels for the predators—it prejudices the outcome in favor of population y. The following lemma gives a sufficient condition for competitive exclusion to hold. It is the counterpart of Lemma 6.2.

LEMMA 7.3. *Let (H1) hold. If, in addition, $m_3 \leq m_2$, then $\lim_{t \to \infty} z(t) = 0$.*

As noted above, when a predator becomes extinct the omega limit set is just the trajectories of the food chain considered in §4. The behavior is summarized as follows:

i) If $\lambda_1 + \lambda_2 > 1$, then $\lim_{t \to \infty} x(t) = 1 - \lambda_1$ and $\lim_{t \to \infty} y(t) = 0$.

ii) If $\lambda_1 + \lambda_2 < 1$, there is an interior critical point (x_c, y_c). If

(7.4) $$\frac{y_c}{m_2 x_c^2} < \frac{m_1 a_1}{(1 + a_1 - x_c - y_c)^2},$$

this critical point is a globally asymptotically stable solution of (7.3). If the inequality (7.4) is reversed, the critical point is unstable and there is at least one periodic orbit.

We now come to a technical difficulty—there may be more than one periodic orbit. (Numerical simulation does not support this possibility.) If there is more than one orbit, the inner one must be stable from the inside and the outer one must be stable from the outside. The first assertion follows from the instability of the critical point and the second from the boundedness of solutions (and, of course, the Poincaré–Bendixson theorem). By a theorem of Erle [19]—or just by counting stability changes (the system is real analytic)—there must be an asymptotically stable limit cycle. One Floquet multiplier of the linearization about this orbit is equal to one and the other must be less than or equal to one for a range in the parameter space. The technical difficulty is that it could be identically one over all of the relevant portion of the parameter space. Yet to apply Theorem 5.1 it will be necessary to have it strictly less than one. Therefore we assume

(H2) There exists a limit cycle of

(7.5) $$x' = x\left(\frac{m_1(1 - x - y)}{1 + a_1 - x - y} - 1 - \frac{m_2 y}{a_2 + x}\right),$$

$$y' = y\left(\frac{m_2 x}{a_2 + x} - 1\right)$$

which has a Floquet multiplier strictly inside the unit circle.

The principal theorem below could be improved by strengthening the bifurcation theorem or by eliminating the possibility of the constant behavior of the Floquet multiplier. It would be a worthwhile question to resolve. We note also that the uniqueness proof of Cheng, used in the previous section, does not appear to work in this case, and uniqueness of the limit cycle remains an open question. Uniqueness of limit cycles is a delicate problem, as shown by Erle [20].

THEOREM 7.4. *Let a_i and m_i be fixed, with $m_i > 1$, $\lambda_i < 1$, $i = 1, 2$, and let $\lambda_1 + \lambda_2 < 1$. Let the coordinates of the resulting critical point satisfy*

$$\frac{y_c}{m_2 x_c^2} > \frac{m_1 a_1}{(1 + a_1 - x_c - y_c)^2}$$

and let (H2) hold. Fix $m_3 > m_2$. Then there exists a number $A > a_2$ such that for $a_3 < A$, $A - a_3$ sufficiently small, $\lambda_2 \leq \lambda_3$ and (7.3) has a periodic orbit in the positive octant arbitrarily near the (x, y)-plane.

Proof. The proof follows the outline given in the preceeding two sections. Let $(x(t), y(t))$ be the asymptotically stable orbit given by (H2). The linearization about this orbit has Floquet multipliers 1 and ν with $|\nu| < 1$. (Actually ν is real and positive.) Then $(x(t), y(t), 0)$ is a periodic orbit of (7.3). Linearize about this solution to obtain a variational matrix of the form

$$\begin{bmatrix} \dfrac{m_1(1-x-y)}{1+a_1-x-y} - 1 - \dfrac{m_2 y}{a_2+x} + \dfrac{m_1 a_1 x}{(1+a_1-x-y)^2} + \dfrac{m_2 xy}{(a_2+x)^2} & \dfrac{-m_2 x}{a_2+x} & \dfrac{-m_3 x}{a_3+x} \\ \dfrac{m_2 a_2 y}{(a_2+x)^2} & \dfrac{m_2 x}{a_2+x} - 1 & 0 \\ 0 & 0 & \dfrac{m_3 x}{a_3+x} - 1 \end{bmatrix}.$$

Let $\mu = \exp\left(\int_0^T [m_3 x(t)/(a_3 + x(t))]\, dt - T\right)$. μ and ν are the eigenvalues of the linearization of the Poincaré map with the origin at this periodic orbit. Note that $x(t)$ is independent of m_3 and a_3. Furthermore,

$$\frac{d\mu}{da_3} < 0.$$

Theorem 5.1 applies to yield a curve of fixed point of the Poincaré map and hence a family of periodic orbits of (7.3) parametrized by a_3.

Questions of global bifurcation and stability are open. However, with an additional hypothesis, Keener [44] has shown that the bifurcation continues. Schematically, things look as in Fig. 6.3 in the (x, y, z)-space. The results of this section should be experimentally verifiable since all of the parameters are measurable in the laboratory.

CHAPTER 3

Some Deterministic Problems in Genetics

1. Introduction. The previous chapter considered interacting populations, based only on ecological considerations. We now wish to expand that viewpoint and add a genetic component. The nature of the problems will be somewhat different, as the focus will be more upon the genetics. However, the key point in two of these problems is the presence of both a genetic and an environmental component. Once the models are developed, the questions and the techniques will be much the same as before—determining the asymptotic behavior of solutions of a system of differential equations by analyzing limit sets.

The usual setting for problems in genetics is with difference equations rather than differential equations, where elementary probability arguments can be used. Difference equations are appropriate in ecology for problems of nonoverlapping generations—insects which lay eggs that hatch after the adults have died—or for populations with a short adult lifespan. Such organisms are frequently the experimental tools of geneticists, and so much of the literature concerns models appropriate for them. However, when the organisms reproduce continuously, there are some conceptual difficulties with the discrete approach, and differential equation formulation seems more appropriate. Moreover, evolution has both a genetic and an environmental component, and if the two are to be integrated, it seems more likely to be done in the continuous formulation since the environmental factors act continuously.

This section presents a general derivation of differential equations describing the evolution of genotypes due to Nagylaki and Crow [52]. The generality will not be emphasized, but specific situations to be encountered in later sections will be described. The reader who is not interested in the derivations may skip ahead to §2. Section 2 considers a predator populations which feeds differently on different genotypes. Section 3 presents a model of cystic fibrosis, a disease which is believed to be the result of a gene at a single locus. Section 4 presents a parental selection model, an example of a "nonstandard" genetics problem in which "steady states" may have unusual symmetries.

The basic model considers a population of one sex—or more correctly, where no genetic properties which distinguish the parents are used. Although our interest will be in the case of two alleles, the model allows any finite number. Let the alleles be labeled by A_i, $i = 1, \cdots, p$, and let there be $2n_i$ of each and a total of N diploid organisms. The number of ordered genotypes $A_i A_j$ is n_{ij} (the number of unordered genotypes is $2n_{ij}$ if $i \neq j$, or n_{ij} if $i = j$). Using ordered genotypes eliminates some constant fractions and makes the derivation symmetric, although we will eventually change to somewhat different notation. The frequency of allele A_i is $p_i = n_i/N$ and the frequency of $A_i A_j$ is $p_{ij} = n_{ij}/N$.

Let d_{ij} be the probability of death of the genotype per unit time, M the number of matings per unit time, $X_{ik,lj}$ the fraction of these which is between genotype A_iA_k and A_lA_j, and $\tilde{a}_{ik,lj}$ the number of progeny from a single $A_iA_k \times A_lA_j$ union. The rate of change of n_{ij} is

$$n'_{ij}(t) = M \sum_{kl} X_{ik,lj} \tilde{a}_{ik,lj} - d_{ij} N p_{ij}.$$

If $a_{ik,lj} = M\tilde{a}_{ik,lj}/N$ then the equation becomes

(1.1) $$n'_{ij} = N \sum_{kl} X_{ik,lj} a_{ik,lj} - d_{ij} n_{ij}.$$

The quantities X,a,d may be functions of time, genotype frequencies, total population, or other relevant parameters, but many simplifying assumptions must be made to even approach a tractable problem. We proceed to use (1.1) to obtain a system of differential equations for the $n_{ij}(t)$ which govern the evolution of the population.

First, there are certain reasonable biological symmetries, for example, $X_{ik,lj} = X_{lj,ik}$, $d_{ij} = d_{ji}$, and $a_{ik,lj} = a_{lj,ik}$. The hypothesis of random mating states that

$$X_{ik,lj} = p_{ik} p_{lj},$$

or that the fraction of matings between A_iA_k and A_lA_j is the product of their frequencies. With this assumption (1.1) becomes (replacing the p_{ij}'s)

(1.2) $$n'_{ij} = \sum_{kl} \frac{n_{ik} n_{lj}}{N} a_{ik,lj} - d_{ij} n_{ij},$$

which will be the basic equation governing the evolution of genotypes.

If $i,j = 1,2$, $a_{ik,lj} = B(N)$, $d_{ij} = \Delta(N)/N$, then (1.2) becomes

$$n'_{11} = \frac{1}{N}[n_{11}^2 + 2n_{12}n_{11} + n_{12}^2]B(N) - n_{11}\frac{\Delta(N)}{N},$$

$$n'_{12} = \frac{1}{N}[n_{11}n_{12} + n_{11}n_{22} + n_{12}^2 + n_{12}n_{22}]B(N) - n_{12}\frac{\Delta(N)}{N},$$

(1.3)

$$n'_{21} = \frac{1}{N}[n_{21}n_{11} + n_{21}^2 + n_{22}n_{11} + n_{22}n_{21}]B(N) - n_{21}\frac{\Delta(N)}{N},$$

$$n'_{22} = \frac{1}{N}[n_{22}^2 + 2n_{12}n_{22} + n_{12}^2]B(N) - n_{13}\frac{\Delta(N)}{N}.$$

Since the coefficients for the equations for $n_{12}(t)$ and $n_{21}(t)$ and are the same, if $n_{12}(0) = n_{21}(0)$, it follows that $n_{12}(t) = n_{21}(t)$ for all t. If one writes $x_1 = n_{11}$, $x_2 = n_{12} + n_{21} = 2n_{12}$, $x_3 = n_{22}$ and $x = x_1 + x_2 + x_3 = N$, (1.3) becomes

$$x'_1 = \frac{(x_1 + \frac{1}{2}x_2)^2}{x} B(x) - \frac{x_1}{x} \Delta(x),$$

(1.4) $$x'_2 = 2\frac{(x_1 + \frac{1}{2}x_2)(x_3 + \frac{1}{2}x_2)}{x} B(x) - \frac{x_2}{x} \Delta(x),$$

$$x'_3 = \frac{(x_3 + \frac{1}{2}x_2)^2}{x} B(x) - \frac{x_3}{x} \Delta(x),$$

SOME DETERMINISTIC PROBLEMS IN GENETICS 49

which is the starting equation used in §2 of this chapter. It describes the evolution of a population with three genotypes, density dependent birth and death, and no selection among genotypes.

In (1.2) suppose $i, j = 1, 2$ and $d_{ij} = 0$. Then (1.2) becomes

$$n'_{11} = \frac{a_{11,11}n_{11}^2 + a_{11,21}n_{11}n_{21} + a_{12,11}n_{12}n_{11} + a_{12,21}n_{12}n_{21}}{N},$$

$$n'_{12} = \frac{a_{11,12}n_{11}n_{12} + a_{11,22}n_{11}n_{22} + a_{12,12}n_{12}^2 + a_{12,22}n_{12}n_{22}}{N},$$

$$n'_{21} = \frac{a_{21,11}n_{21}n_{11} + a_{21,21}n_{21}^2 + a_{22,11}n_{22}n_{11} + a_{22,21}n_{22,21}}{N},$$

$$n'_{22} = \frac{a_{21,12}n_{21}n_{12} + a_{21,22}n_{21}n_{22} + a_{22,12}n_{22}n_{12} + a_{22,22}n_{22}^2}{N}.$$

If one assumes $a_{12,kl} = a_{21,kl}$ and a $a_{kl,12} = a_{kl,21}$, then there are only three subscripts, 11, 12, or 22, which can be written as 1, 2, or 3, respectively. Using B's with two subscripts instead of a's with four and letting $x_1 = n_{11}$, $x_2 = n_{12} + n_{21} = 2n_{21}$ (if $n_{12}(0) = n_{21}(0)$) and $x_3 = n_{22}$, the above system simplifies to

$$x'_1 = \frac{1}{x}\left[B_{11}x_1^2 + B_{12}x_1x_2 + \frac{B_{22}}{4}x_2^2\right],$$

(1.5) $$x'_2 = \frac{2}{x}\left[B_{13}x_1x_3 + \frac{B_{12}}{2}x_1x_2 + \frac{B_{23}}{2}x_2x_3 + \frac{B_{22}}{4}x_2^2\right],$$

$$x'_3 = \frac{1}{x}\left[B_{33}x_3^2 + B_{23}x_2x_3 + \frac{B_{22}}{4}x_2^2\right].$$

The system (1.5) reflects a parental selection model where the genotype of the parents determines the fecundity. This is usually interpreted in terms of increased egg laying capacity, although [28] also cites evidence in humans. The system (1.5) will be analyzed in §4 of this chapter.

When separate sexes are considered, the same general principles apply. Again, let M be the number of matings per time unit, $X_{ik,lj}$ the proportion of these between A_iA_k males and A_lA_j females, $\tilde{a}_{ik,lj}$ and $\bar{\tilde{a}}_{ik,lj}$ the number of male and female progeny from a single such union. Then if $\bar{n}_{ij}$ is the number of females of type A_iA_j and n_{ij} the number of males, the respective rates of change are

(1.6)
$$n'_{ij} = \frac{1}{2}M\sum_{kl}(X_{ik,lj}\tilde{a}_{ik,lj} + X_{lj,ik}\tilde{a}_{lj,ik}) - d_{ij}NP_{ij},$$

$$\bar{n}'_{ij} = \frac{1}{2}M\sum_{kl}(X_{ik,lj}\bar{\tilde{a}}_{ik,lj} + X_{lj,ik}\bar{\tilde{a}}_{lj,ik}) - \bar{d}_{ij}\bar{N}P_{ij}$$

where the barred quantities refer to females; the ½ represents the split of sexes and lets (1.6) add up to (1.2) in the case where all quantities are the same, by attributing half "credit" for each child to each parent. Let

$$a_{ik,lj} = \frac{M\tilde{a}_{ik,lj}}{N}, \qquad \bar{a}_{ik,lj} = \frac{M\bar{\tilde{a}}_{ik,lj}}{\bar{N}}$$

and assume random mating,

$$X_{ij,kl} = \frac{n_{ij}}{N} \frac{\bar{n}_{kl}}{N}.$$

Then the system of differential equations for the evolution of the population becomes

(1.7)
$$n'_{ij} = \frac{1}{2} \sum_{kl} \frac{[\bar{n}_{ik} n_{lj} a_{ik,lj} + \bar{n}_{lj} n_{ik} a_{lj,ik}]}{N} - d_{ij} n_{ij},$$

$$\bar{n}'_{ij} = \frac{1}{2} \sum_{kl} \frac{[\bar{n}_{ik} n_{lj} \bar{a}_{ik,lj} + \bar{n}_{lj} n_{ik} \bar{a}_{lj,ik}]}{N} - \bar{d}_{ij} \bar{n}_{ij},$$

This is the counterpart of (1.2).

In the system (1.7), if one assumes that the sex of the offspring is not affected, then $a_{ik,lj} = \bar{a}_{ik,lj}$. If, further, the death rate is independent of sex, $d_{ij} = \bar{d}_{ij}$. Then (1.7) becomes

$$n'_{ij} = \frac{1}{2N} \sum_{kl} n_{ik} \bar{n}_{lj} (a_{ik,lj} + a_{lj,ik}) - d_{ij} n_{ij},$$

$$\bar{n}'_{ij} = \frac{1}{2N} \sum_{kl} \bar{n}_{ik} n_{lj} (a_{ik,lj} + a_{lj,ik}) - d_{ij} \bar{n}_{ij}.$$

It is reasonable to assume that $a_{ij,kl} = a_{ij,lk}$, and $a_{ij,kl} = a_{ji,kl}$, that is, that only the genotype is important, not the parent from which the gene is inherited. Then it is not hard to show that if $n_{ij}(0) = \bar{n}_{ij}(0)$, $n_{ij}(t) = \bar{n}_{ij}(t)$ for all t. One can then add the equations and reduce the number of variables. For the two-allele case, then, there are only three "pairs"—11, 12, 22—so the subscripting can be simplified by denoting $a_{11,11}$ as B_{11}, $a_{12,11}$ as B_{21}, etc. For this case one has then the following equations:

(1.8)
$$n'_{11} = \frac{1}{N} [2B_{11} n^2_{11} + (B_{12} + B_{21}) n_{11} n_{21}$$

$$+ (B_{21} + B_{12}) n_{12} n_{11} + 2B_{22} n_{12} n_{21}] - d_{11} n_{11},$$

$$n'_{12} = \frac{1}{N} [(B_{12} + B_{21}) n_{11} n_{12} + (B_{13} + B_{31}) n_{11} n_{22}$$

$$+ 2B_{22} n^2_{12} + (B_{32} + B_{23}) n_{12} n_{22}] - d_{12} n_{12},$$

$$n'_{21} = \frac{1}{N} [(B_{21} + B_{12}) n_{21} n_{11} + 2B_{22} n^2_{21}$$

$$+ (B_{31} + B_{13}) n_{22} n_{11} + (B_{23} + B_{32}) n_{22} n_{21}] - d_{21} n_{21},$$

$$n'_{22} = \frac{1}{N} [2B_{22} n_{21} n_{12} + (B_{23} + B_{32}) n_{21} n_{22}$$

$$+ (B_{32} + B_{23}) n_{22} n_{12} + 2B_{33} n^2_{22}] - d_{22} n_{22}.$$

If $d_{21} = d_{12}$, the equations for n_{12} and n_{21} are the same, so $n_{21}(0) = n_{12}(0)$ implies that $n_{21}(t) = n_{12}(t)$ for all t. Letting $x_1 = n_{11}$, $x_2 = n_{12} + n_{21} = 2n_{12}$ and $x_3 = n_{22}$ and denoting d_{11} as d_1, d_{12} as d_2, and d_{22} as d_3 gives the system

(1.9)
$$x_1' = \frac{1}{x}\left[2B_{11}x_1^2 + (B_{12} + B_{21})x_1x_2 + \frac{B_{22}}{2}x_2^2\right] - d_1x_1,$$

$$x_2' = \frac{1}{x}[(B_{12} + B_{21})x_1x_2 + 2(B_{13} + B_{31})x_1x_3$$
$$+ B_{22}x_2^2 + (B_{23} + B_{32})x_2x_3] - d_2x_2,$$

$$x_3' = \frac{1}{x}\left[2B_{33}x_3^2 + (B_{32} + B_{23})x_3x_2 + \frac{B_{22}}{2}x_2^2\right] - d_3x_3.$$

Sterility can be modeled by setting the appropriate B equal to zero; for example, if heterozygotes cannot produce offspring, one could set $B_{22} = 0$ in (1.9). The system modeling cystic fibrosis in §3 will be derived from (1.9).

2. Predator influence on the growth of a population with three genotypes. Earlier, a model of a population consisting of three genotypes was derived. The variables x_1, x_2, x_3 correspond to the three possible combinations of two alleles at a single locus in a diploid organism. The equations took the form

(2.1)
$$x_1' = \frac{(x_1 + \tfrac{1}{2}x_2)^2}{x}B(x) - \frac{x_1}{x}\Delta(x),$$

$$x_2' = \frac{2(x_1 + \tfrac{1}{2}x_2)(x_3 + \tfrac{1}{2}x_2)}{x}B(x) - \frac{x_2}{x}\Delta(x),$$

$$x_3' = \frac{(x_3 + \tfrac{1}{2}x_2)^2}{x}B(x) - \frac{x_3}{x}\Delta(x),$$

$$x = x_1 + x_2 + x_3, \qquad x_i(0) = x_{i0} \geq 0.$$

The total population x satisfies

$$x' = xB(x) - \Delta(x),$$

a typical growth equation.

To see the genetic consequences of this model, first let $\Delta(x) \equiv 0$ and introduce variables u and v (representing half of the total number of each allele)

$$u(t) = x_1(t) + \tfrac{1}{2}x_2(t), \qquad v(t) = x_3(t) + \tfrac{1}{2}x_2(t).$$

The differential equations for u can be derived for

$$u'(t) = x_1'(t) + \tfrac{1}{2}x_2'(t)$$
$$= \frac{(x_1 + \tfrac{1}{2}x_2)^2}{x}B(x) + \frac{(x_1 + \tfrac{1}{2}x_2)(x_3 + \tfrac{1}{2}x_2)}{x}B(x)$$
$$= \frac{u^2}{u+v}B(u+v) + \frac{uv}{u+v}B(u+v) = uB(u+v).$$

Similarly,
$$v'(t) = vB(u+v).$$

Since
$$\frac{du}{dv} = \frac{u'(t)}{v'(t)} = \frac{uB(u+v)}{vB(u+v)},$$

one has the first order equation
$$\frac{du}{dv} = \frac{u}{v}$$

governing the relative growth of u and v. A solution of this simple equation is $u = cv$, where
$$c = \frac{u_0}{v_0} = \frac{x_{10} + \tfrac{1}{2}x_{20}}{x_{30} + \tfrac{1}{2}x_{20}}.$$

Returning to the original variables one has
$$\frac{dx_1}{dx_3} = \frac{u^2}{v^2} = c^2, \qquad \frac{dx_2}{dx_3} = \frac{2uv}{v^2} = 2c.$$

Thus
$$x_1 - x_{10} = c^2(x_3 - x_{30}) \quad \text{and} \quad x_2 - x_{20} = 2c(x_3 - x_{30}).$$

A more standard way to write this result is as a relative proportion:
$$x_1(t) - x_{10} : x_2(t) - x_{20} : x_3(t) - x_{30} = c^2 : 2c : 1.$$

This is the *Hardy–Weinberg* principle in genetics. It says that the number of offspring—$x_i(t) - x_{i0}$, $i = 1, 2, 3,$—of each type are in the same relative proportions, $c^2 : 2c : 1$, for all $t > 0$ where c, given above, is determined by the initial ratio of alleles.

If a nontrivial, nongenetically related, external death rate is assumed ($\Delta(x) > 0$ in the above discussion), the equations for u and v become
$$u' = uB(u+v) - \frac{u}{u+v}\Delta(u+v), \qquad v' = vB(u+v) - \frac{v}{u+v}\Delta(u+v)$$

which yields the same first order equation as before. However, the remaining analysis is slightly more delicate. We consider the quantity $z = x_1(t) - c^2 x_3(t)$ (anticipating a similar result) and find that
$$z' = x_1'(t) - c^2 x_3'(t) = \frac{u^2}{x}B(x) - \frac{x_1}{x}\Delta(x) - \frac{c^2 v^2}{x}B(x) - c^2\frac{x_3}{x}\Delta(x)$$
$$= (x_1 - c^2 x_3)\frac{\Delta(x)}{x} = -z\frac{\Delta(x)}{x}.$$

Hence
$$x_1(t) - c^2 x_3(t) = [x_1(0) - c^2 x_3(0)] \exp\left[\int_0^t \frac{\Delta(x(s))}{x(s)} ds\right].$$

If the integral should diverge (which is interpreted that death is certain) then $\lim_{t \to \infty} [x_1(t) - c^2 x_3(t)] = 0$. Similar arguments can be made to yield $\lim_{t \to \infty} [x_2(t) - 2c\, x_3(t)] = 0$ or that the subpopulations are *asymptotic* to Hardy–Weinberg proportions (as expected since now the initial population is subject to the death rate).

For a population which exhibits logistic growth, one could, for example, take $B(x) \equiv \alpha$, α constant, and $\Delta(x) = \alpha x^2/K$, K constant. The system would take the form

$$x_1' = \frac{(x_1 + \tfrac{1}{2} x_2)^2}{x} \alpha - \frac{\alpha}{K} x_1 x,$$

$$x_2' = \frac{2(x_1 + \tfrac{1}{2} x_2)(x_3 + \tfrac{1}{2} x_2)}{x} \alpha - \frac{\alpha}{K} x_2 x,$$

$$x_3' = \frac{(x_3 + \tfrac{1}{2} x_2)^2}{x} \alpha - \frac{\alpha}{K} x_3 x.$$

A more interesting question is to consider what happens if the three offspring are not exactly the same. This can occur in many ways; for example, one gene can confer added reproductive ability or added survival ability. Such attributes are grouped under a general heading of "fitness." We consider one simple problem, involving a predator. The material follows Freedman and Waltman [24].

Denote the two alleles by A and a. Let one of the alleles, A, be dominant. Then AA and Aa will appear to be the same—they are said to be of the same phenotype. An example is that brown eyes dominate blue—an individual has blue eyes only if he carries both markers for blue—aa. Suppose that an allele for one of two possible colors of an organism is dominant, the other, recessive. Color may affect the susceptibility to predation, and one wishes to modify (2.1) to reflect this. First, however, one must specify the predation interaction. The simplest model is to accept the Lotka–Volterra (or mass action) formulation that predation is proportional to contacts, i.e., of form ρxy where ρ is constant, x is the prey, and y the predator. One can also adopt the Michaelis–Menten formulation as a predator–prey response, $Pxy/(a + x)$. As noted before, this nonlinearity does have a basis as a predator–prey response, usually associated with the name of Holling. It reflects the fact that a relatively small proportion of the prey are being pursued at any one time. Here we take a sufficiently general function to include both of these important special cases, but the hypothesis of the theorem will place a further restriction by requiring the existence of a certain asymptotically stable, interior critical point. We assume that interactions between predator and prey are of the form $yP(x, y)$, with $P(0, y) = 0$ and $P_x(x, y) > 0$. Since, in fact, there are three prey subpopulations, we assume that the interactions take the form

$$\frac{x_i y}{x} P_i(x, y), \quad i = 1, 2, 3, \quad x = x_1 + x_2 + x_3.$$

The modified equations become

$$x'_1 = \frac{(x_1 + \tfrac12 x_2)^2}{x} B(x) - (\Delta(x) + yP_1(x,y))\frac{x_1}{x},$$

$$x'_2 = \frac{2(x_1 + \tfrac12 x_2)(x_3 + \tfrac12 x_2)}{x} B(x) - (\Delta(x) + yP_2(x,y))\frac{x_2}{x},$$

(2.2)

$$x'_3 = \frac{(x_3 + \tfrac12 x_2)^2}{x} B(x) - (\Delta(x) + yP_3(x,y))\frac{x_3}{x},$$

$$y' = y\left(-s + k\sum_{i=1}^{3}\frac{x_i}{x}P_i(x,y)\right),$$

$$x_i(0) = x_{i0} > 0, \quad y(0) = y_0 > 0, \quad x = x_1 + x_2 + x_3.$$

"Logistic-like" growth is modeled in two hypotheses:

(H1) $\Delta(x) \geq 0$, $B(x) > 0$, $x \neq 0$, $\Delta(0) = 0$, $B(0) > \Delta'(0) > 0$.

(H2) There is a unique number $K > 0$ such that $KB(K) = \Delta(K)$, and $KB'(K) + B(K) < \Delta'(K)$.

The above assumptions on the functions P_i can be expressed as

(H3) $\quad P_i(0, y) = 0, \quad P_{ix}(x,y) > 0, \quad P_{iy}(x,y) \geq 0.$

Finally, one adds

(H4) The system

(2.3) $\qquad x' = xB(x) - \Delta(x) - yP_3(x,y), \qquad y' = y(-s + kP_3(x,y))$

has a globally (with respect to the open positive octant) asymptotically stable critical point (x^*, y^*) in the interior of the positive octant.

(Although we are assuming an attractor in the form of a critical point, a more general attractor can be allowed.)

Two special cases illustrate that the condition (H4) is at least sometimes reasonable. If we interpret the $B(x)$ and $\Delta(x)$ as logistic and $P_i(x,y) = p_i x$, p_i a positive constant, $i = 1, 2, 3$, then (2.3) takes the form

(2.4) $\qquad x' = x(1 - x) - p_3 xy, \qquad y' = y(-s + kp_3 x).$

The critical point

$$x^* = \frac{s}{kp_3}, \qquad y^* = \frac{1 - x^*}{p_3} = \frac{kp_3 - s}{kp_3^2}$$

turns out to be globally asymptotically stable.

If the logistic growth is retained for the predator and Michaelis–Menten dynamics is used for predation, the system (2.3) becomes

SOME DETERMINISTIC PROBLEMS IN GENETICS

(2.5)
$$x' = x(1 - x) - \frac{p_3 xy}{a_3 + x},$$

$$y' = y\left(-s + k\frac{p_3 x}{a_3 + x}\right).$$

The interior critical point is $x^* = sa_3/(kp_3 - s)$, $y^* = (1 - x^*)(a_3 + x^*)/p_3$, provided $0 < x^* < 1$. That this critical point is logically asymptotically stable with respect to the positive octant was noted in Chapter 2.

Finally it is necessary to make a precise statement of what it means for the predation function for one genotype to be larger than for the other. Recall that we have in mind that A is dominant so that x_1 and x_2 appear the same to a predator and are more vulnerable to predation than x_3, which corresponds to the total lack of the allele A. This is expressed as

(H5) $P_1(x, y) = P_2(x, y) \geq 0$, $\displaystyle\inf_{\substack{K \geq x > 0 \\ M \geq y > 0}} \frac{P_1(x, y) - P_3(x, y)}{x} \geq \delta(M) > 0$,

where K is given in (H2).

The problem is not of interest unless the predator can "live" on the prey in the first place. We say that the predator can *survive* (on the prey) of $\limsup_{t \to \infty} y(t) > 0$. The mathematical result is stated in the following theorem.

THEOREM 2.1. *Let* (H1)–(H5) *hold and suppose that the predator survives. Then solutions of* (2.2) *satisfy*

$$\lim_{t \to \infty} x_i(t) = 0, \quad i = 1, 2, \quad \lim_{t \to \infty} x_3(t) = x_3^* > 0, \quad \lim_{t \to \infty} y(t) = y^* > 0$$

where (x_3^*, y^*) *are the coordinates of the critical point given in* (H4).

The theorem states that organisms carrying the A allele become extinct. If the pairs $(A, A)(A, a)$ correspond, say, to light coloring in some form of insect and (a, a) to dark coloring, then in a rural environment the two genotypes may be equal in their susceptibility to predation, while in an industrial environment—a "dirty" environment—being dark might give added protection from predation. If the model should apply with (H1)–(H4) and the advantage of being dark is as strong as (H5) then only x_3—the carrier of the recessive gene—survives. (Actually the dominant and recessive roles are irrelevant and the inequality in (H5) may be reversed if (H4) is appropriately modified.)

The proof of the theorem involves several small lemmas in which (H1)–(H5) are assumed. Since similar results are needed in the next section, proofs will be given.

LEMMA 2.2. *The positive cone is invariant. Solutions with initial conditions in the positive cone are bounded.*

Proof. First of all, the equation for y may be written in integral form,

$$y(t) = y(0) \exp \int_0^t \left(-s + k \sum_{i=1}^{3} \frac{x_i}{x} P_i(x, y)\right) dt,$$

so that if $y(0) > 0$, $y(t) > 0$ for all $t > 0$. Furthermore, each $x_i(t)$ is a solution of the inequality

$$x_1'(t) > -(\Delta(x) + yP_i(x, y))\frac{x_i}{x},$$

so that, if $x_i(0) > 0$, then

$$x_i(t) > x_i(0) \exp\left(-\int_0^t \left[\Delta(x(t)) + \frac{yP_i x(t), y(t)}{x(t)}\right] dt\right)$$

and $x_i(t) > 0$ for $t > 0$. (Note that $x_i(0) > 0$ means $x(0) > 0$ and hence that $B(x) > 0$ near $x = 0$, so that the inequality is strict.)

Since

$$x'(t) \le xB(x) - \Delta(x), \qquad x(0) = x_0,$$

it follows by comparison that $0 < x(t) \le K + \varepsilon$, for $\varepsilon > 0$ and t sufficiently large, which bounds $x(t)$, and since each $x_i(t)$ is positive, bounds $x_i(t)$. There are constants a and R such that

$$(ax + y)' \le -s(ax + y) + R$$

for t large. Since $ax + y$ is then bounded by the same comparison theorem, and $ax > 0$, it follows that $y(t)$ is bounded. This result is, of course, intuitively clear—if the prey population is bounded, the predator population must be also.

LEMMA 2.3. *There are no critical points of the system* (2.2) *which are interior to the positive cone.*

Proof. Suppose, to the contrary, that $P = (\tilde{x}_1, \tilde{x}_2, \tilde{x}_3, \tilde{y})$ were such a critical point. Let

$$u(t) = x_1(t) + \tfrac{1}{2} x_2(t), \qquad v(t) = x_3(t) + \tfrac{1}{2} x_2(t)$$

and observe that $u'(t)$ and $v'(t)$ take the form

$$u' = \frac{u}{x}(xB(x) - \Delta(x)) - \frac{y}{x} uP_1(x, y),$$

$$v' = \frac{v}{x}(xB(x) - \Delta(x)) - \frac{y}{x}\left[x_3 P_3(x, y) + \frac{x_2}{2} P_1(x, y)\right].$$

At a critical point, $x_1' = x_2' = x_3' = 0$, so $u' = v' = 0$. Let $\tilde{u}, \tilde{v}$ correspond to $\tilde{x}_1, \tilde{x}_2, \tilde{x}_3$. Note that $\tilde{u} > 0, \tilde{v} > 0$ since $\tilde{x}_i > 0$ by hypothesis. Then

$$\frac{\tilde{u}}{\tilde{v}} = \frac{\tilde{u} P_1(\tilde{x}, \tilde{y})}{\tilde{x}_3 P_3(\tilde{x}, \tilde{y}) + (\tilde{x}_2/2) P_1(\tilde{x}, \tilde{y})} > \frac{\tilde{u}}{\tilde{v}}$$

since $P_3(\tilde{x}, \tilde{y}) < P_1(\tilde{x}, \tilde{y})$. This contradiction establishes the lemma.

LEMMA 2.4. *Let* $(x_1(t), x_2(t), x_3(t), y(t))$ *be a solution of the system* (2.2). *If* $\int_0^\infty x_3(t) y(t) dt = \infty$, *the conclusion of Theorem* 2.1 *holds.*

Proof. Let $u(t)$ and $v(t)$ be as in Lemma 2.3. Then one has that

$$\frac{u'}{u} - \frac{v'}{v} = \frac{x_3 y}{xv} [P_3(x, y) - P_1(x, y)]$$

or that

(2.6) $$u(t) \leq cv(t) \exp\left(-\frac{\delta}{K}\int_0^t x_3(s)y(s)\,ds\right)$$

where $\delta = \delta(M)$ and M is the bound on $y(t)$ given in Lemma 2.2. Since $v(t)$ is a bounded function it follows that $\lim_{t\to\infty} u(t) = 0$. This is possible only if $\lim_{t\to\infty} x_1(t) = 0$ and $\lim_{t\to\infty} x_2(t) = 0$. Thus there is a point in the omega limit set of the form $(0, 0, \tilde{x}_3, \tilde{y})$. If $\tilde{x}_3 > 0$ and $\tilde{y} > 0$, then the trajectory through this point is a solution of

(2.7) $\quad x_3' = x_3 B(x_3) - \Delta(x_3) - yP_3(x_3, y), \qquad y' = y(-s + P_3(x_3, y)).$

However, by hypothesis (H4) this system has a globally asymptotically stable critical point which must then be the only omega limit point—therefore $\tilde{x}_3 = x^*$, $\tilde{y} = y^*$ where (x_3^*, y^*) are given in (H4). If $\tilde{x}_3 = 0$, then the solution of (2.7) through $(0, \tilde{y})$ satisfies $\lim_{t\to\infty} y(t) = 0$, contrary to our hypothesis that the predator survives.

LEMMA 2.5. *No solution of* (2.2) *with initial conditions in the positive cone has an omega limit point of the form* $(\tilde{x}_1, \tilde{x}_2, 0, \tilde{y})$ *with* $\tilde{x}_2 > 0$.

Proof. As noted in Lemma 2.2, any solution with initial conditions in the positive cone must remain there for all later times. Let $\Gamma(t)$ be a trajectory with $\Gamma(0)$ in the positive cone and with an omega limit point of the form $P = (\tilde{x}_1, \tilde{x}_2, 0, \tilde{y})$. The entire trajectory through this point must also be in the omega limit set. Call the trajectory through P at $t = 0$ $\gamma(t)$, $\gamma(t) = (\bar{x}_1(t), \bar{x}_2(t), \bar{x}_3(t), \bar{y}(t))$. Since

$$\bar{x}_3'\Big|_{t=0} = \frac{\tilde{x}_2^2 B(\tilde{x})}{(\tilde{x}_1 + \tilde{x}_2)} > 0,$$

$x_3(t) < 0$ for t in an interval $(-a, 0)$, $a > 0$. Let t_n, $t_n \to \infty$ denote the sequence such that $\Gamma(t_n) \to P$. Then $\Gamma(t_n - \delta) \to \gamma(-\delta)$ as $t_n \to \infty$. Thus $\Gamma(t_n - \delta)$ is not in the positive cone for n large and $0 < \delta < a$, contradicting Lemma 2.2.

Proof of Theorem 2.1. Lemma 2.4 completes the proof in case a trajectory of (2.2) satisfies $\int_0^\infty x_3(t)y(t)\,dt = \infty$. Suppose $\Gamma(t) = (x_1(t), x_2(t), x_3(t), y(t))$ is a trajectory of (2.2), with initial conditions in the positive cone, such that $\int_0^\infty x_3(t)y(t)\,dt < \infty$. $(x_3(t)y(t))'$ is bounded for $t > 0$ since the trajectory is bounded and (2.2) can be used to write this derivative as a combination of bounded terms. Thus $x_3(t)y(t)$ is a uniformly bounded positive function with $\int_0^\infty x_3(t)y(t)\,dt$ finite and so

$$\lim_{t\to\infty} x_3(t)y(t) = 0.$$

Since $\limsup_{t\to\infty} y(t) = \alpha > 0$, either (i) $\lim_{t\to\infty} y(t) = \alpha$ or (ii) there is a sequence $t_n \to \infty$, such that $y'(t_n) = 0$ and $\lim_{n\to\infty} y(t_n) = \alpha$. In the former case, $\lim_{t\to\infty} x_3(t) = 0$, so that by Lemma 2.5, $\lim_{t\to\infty} x_2(t) = 0$ and by (2.6), $\lim_{t\to\infty} x_1(t) = 0$. Since $\lim_{t\to\infty} x(t) = 0$, for t sufficiently large $y'(t) < -sy(t)/2$ which leads to a contradiction. In case (ii), $\lim_{n\to\infty} x_3(t_n) = 0$, so $\lim_{n\to\infty} x_2(t_n) = 0$ by Lemma 2.5 and

$\lim_{n\to\infty} x_1(t_n) = 0$ by (2.6). On the other hand, at t_n,

$$\sum_{1}^{3} \frac{x_i(t_n)}{x(t_n)} P_i(x(t_n), y(t_n)) = \frac{s}{k} \quad \text{or} \quad P_1(x(t_n), y(t_n)) > \frac{s}{k}.$$

This and (H1) imply there is a subsequence t'_n such that $x(t'_n) \to Q > 0$, which is incompatible with $\lim_{n\to\infty} x(t_n) = 0$. Thus it must be the case that $\int_0^\infty x_3(t)y(t)dt = \infty$, and the proof is complete.

3. A model of cystic fibrosis. Cystic fibrosis is a disease which is believed to be caused by a gene at a single locus [17] and which occurs in roughly one of two thousand live births. The disease results in a reduced lifespan and males that do reach maturity are sterile. How then can the disease persist in the population at such a high rate? A possible explanation is that the heterozygote has an advantage—some trait which leads to an increase in the amount of the gene in the general population. A study [2] of the grandparents of cystic fibrosis patients found that as a group they had a higher number of offspring—4.67 compared to 4.03. (Grandparents were used rather than parents because they were unaware that they were carriers of the disease and hence their family size was unaffected.) The cause of the increased progeny is unknown. K. Beck [5] modeled this disease using some of the ideas in the previous section. This section describes her model and the analysis based on it. The notation, proofs and figures are all from this work.

It was noted in Chapter 1 that the census data in the United States did not track as the solution of a logistic equation. The first problem in constructing the model, if one wants it to be predictive, is to model the population growth of the United States.

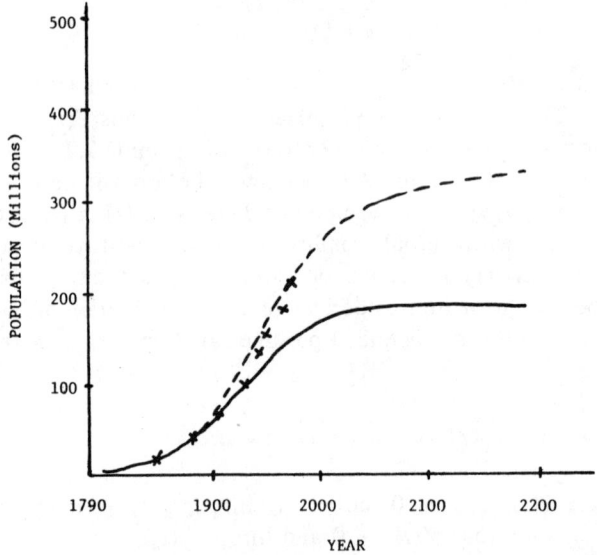

FIG 3.1. *The system* (3.1) *fitted to the census data (dashed line) compared with a straight logistic fit. From Beck* [5]. [Copyright 1982, Elsevier Science Publishing Co. Used by permission.]

SOME DETERMINISTIC PROBLEMS IN GENETICS

Although the logistic equation with constant carrying capacity does not fit, it may fail because the "carrying capacity" is changing—perhaps improving with technology. Beck's approach was to use

(3.1)
$$x' = rx(t)\left(1 - \frac{x(t)}{K(t)}\right), \quad K'(t) = BK(t)\left(1 - \frac{K(t)}{L}\right),$$
$$x(0) = x_0, \quad K(0) = K_0$$

which says that the carrying capacity itself evolves logistically. There are many other possibilities, but (3.1) fits well—see Fig. 3.1 where $r = .0322$, $B = .008$, $L = 350$, and $K(0) = 184$ (the zero year being 1790). Of course, the equation for K could be solved and the result used to write only one equation in (3.1). Although (3.1) is used in the simulations, the mathematics is more general, and arbitrary birth, death and growth functions will be used.

Let $x_1(t)$ denote the number of individuals who do not carry the cystic fibrosis gene at time t, $x_2(t)$ the number of heterozygotes at time t, and $x_3(t)$ the number of individuals with cystic fibrosis at time t. If no differences existed with respect to fertility and survival between these genotypes then the system

(3.2)
$$x_1' = \frac{(x_1 + \tfrac{1}{2} x_2)^2}{x} B(x, K) - \frac{x_1}{x}\Delta(x, K),$$

$$x_2' = \frac{2(x_1 + \tfrac{1}{2} x_2)(x_3 + \tfrac{1}{2} x_2)}{x} B(x, K) - \frac{x_2}{x}\Delta(x, K),$$

$$x_3' = \frac{(x_3 + \tfrac{1}{2} x_2)^2}{x} B(x, K) - \frac{x_3}{x}\Delta(x, K),$$

$$K' = G(K) - D(K),$$

$$x_i(0) = x_{i0} \geq 0, \quad i = 1, 2, 3, \quad x = x_1 + x_2 + x_3, \quad K(0) = K_0 \geq 0,$$

would be appropriate. This is (1.4), derived in §1, with a K equation added, and with births and deaths dependent on both K and x.

(3.2) must be modified to account for three properties of the disease:
 (i) Female heterozygotes have more progeny.
 (ii) Individuals with the disease do not reproduce.
 (iii) Individuals with the disease have greatly reduced survival probabilities.

Since the female has different reproductive characteristics, it is necessary to use (1.6) rather than (1.1) as a starting point. We can begin with (1.9), however, since the sex of each offspring is not affected. With a K-equation added, (1.9) takes the form

$$x_1' = \frac{1}{x}\left[2B_{11}x_1^2 + (B_{12} + B_{21})x_1 x_2 + \frac{B_{22}x_2^2}{2}\right] - d_1 x_1,$$

$$x_2' = \frac{1}{x}[(B_{12} + B_{21})x_1 x_2 + 2(B_{13} + B_{31})x_1 x_3 + B_{22}x_2^2 + (B_{32} + B_{23})x_2 x_3] - d_2 x_3,$$

$$x_3' = \frac{1}{x}\left[2B_{33}x_3^2 + (B_{32} + B_{23})x_3x_2 + \frac{B_{22}}{2}x_2^2\right] - d_3x_3,$$

$$K' = G(K) - D(K),$$

where B_{ij} can be functions of K as well as the number of individuals in the population. To model the lack of reproductive capacity, any B with a subscript "3" is set equal to zero—an individual with the disease may not be a parent. The differing death rates may be modeled by letting $d_1 = d_2 = \Delta_1(x, K)/x$ and $d_3 = \Delta_2(x, K)/x$. Finally let $B_{i2} = \alpha B(x, K)/2$ and $B_{i1} = B(x, K)/2$, where $\alpha > 1$ and $i = 1, 2$, and $B(x, K)$ is to be specified. This reflects the assumption that heterozygote mothers have more children and that density limitations are a function of total population size.

The above system now becomes

(3.3)
$$x_1' = \frac{1}{x}\left(x_1 + \frac{\alpha}{2}x_2\right)\left(x_1 + \frac{1}{2}x_2\right)B(x, K) - \frac{x_1}{x}\Delta_1(x, K),$$

$$x_2' = \frac{1}{x}\left(\frac{1}{2}x_1x_2 + \frac{\alpha}{2}x_2x_1 + \frac{\alpha x_2^2}{2}\right)B(x, K) - \frac{x_2}{x}\Delta_1(x, K),$$

$$x_3' = \frac{\alpha}{4}\frac{x_2^2}{x}B(x, K) - \frac{x_3}{x}\Delta_2(x, K),$$

$$K' = G(K) - D(K),$$

$$x_i(0) = x_{i0} \geq 0, \quad i = 1, 2, 3, \quad K(0) = K_0 \geq 0.$$

The following hypotheses are made:

(H1) $\alpha > 1$.

(H2) There is a $\delta(L) > 0$ such that $(\Delta_2(x, K) - \Delta_1(x, K))/x > \delta(L)$ for all $L \geq x > 0$ and $L \geq K > 0$ (L will be fixed in (H5)).

(H3) $\Delta_1(x, K) \geq 0$, $B(0, K) = \Delta_1(0, K) = \Delta_2(0, K) = 0$, $xB(x, K) > \Delta_1(x, K)$ if $0 < x < K$.

(H4) $\alpha xB(x,K) = \Delta_1(x,K)$, $x > 0$, if and only if $x = \alpha K$ and $\alpha xB(x,K) < \Delta_1(x, K)$ if $x > \alpha K$.

(H5) There is a unique positive number L such that $G(L) = D(L)$ and at this point $G'(L) < D'(L)$, $D(0) \geq 0$, $G(0) = D(0) = 0$, $G'(0) > D'(0) \geq 0$.

(H6) $0 < K_0 < L$ and $0 < x_{10} + x_{20} + x_{30} < K_0$, L as in (H5).

(H1) and (H2) reflect properties of the disease while (H3), (H4) and (H5) are standard logistic-like assumptions and (H6) indicates that the population is initially below carrying capacity (a technical assumption needed in the proof). Finally we note that it is reasonable to make $x_3(0)$ small, at least relative to the other initial values—this will be made precise in the hypotheses of the theorem.

THEOREM 3.1. *Let* (H1)–(H6) *hold. If* $x_{10} + \frac{1}{2}x_{20} > 0$, $x_{30} + \frac{1}{2}x_{20} > 0$, $x_{30} \leq x_{10}$, *and* $x_{30}x_{10} - (x_{20})^2 < 0$ *then solutions of* (3.3) *satisfy* $\liminf_{t\to\infty} x_3(t) + \frac{1}{2}x_2(t) > 0$.

Since $x_3(t) + \frac{1}{2}x_2(t)$ represents one-half the number of deleterious genes, the conclusion of the theorem says that the *gene persists in the population*.

By contrast, if $\alpha = 1$ (no advantage to the heterozygote in the number of offspring) but the other hypotheses hold, then the next theorem says that the deleterious gene becomes extinct. This, coupled with the evidence of more children for heterozygote mothers offers an explanation of the presence of the gene in the population today.

THEOREM 3.2. *Under the hypotheses* (H2)–(H6), *if* $x_{10} + \frac{1}{2}x_{20} > 0$ *and* $x_{30} + \frac{1}{2}x_{20} > 0$ *and* $\alpha = 1$, *then* $\lim_{t\to\infty} x_3(t) = \lim_{t\to\infty} x_2(t) = 0$.

Some preliminary lemmas are needed. (H1)–(H6) are assumed as appropriate without further mention.

LEMMA 3.1. *The region* $T = \{(x_1, x_2, x_3, K) \mid 0 < x_1 + x_2 + x_3 \leq \alpha K, K < L, x_i \geq 0, i = 1, 2, 3\}$ *is positively invariant.*

LEMMA 3.2. *There is no omega limit point for* (3.2) *of the form* $(\tilde{x}_1, \tilde{x}_2, 0, \tilde{K})$ *with* $\tilde{x}_2 > 0$.

The proofs are essentially the same as those for similar lemmas in §2.

Proof of Theorem 3.2. Let $u = x_1 + \frac{1}{2}x_2$, $V = x_3 + \frac{1}{2}x_2$. Since $\alpha = 1$,

$$u' = \frac{1}{x} B(x, K)\left(u + \frac{1}{2}x_2\right)u - \frac{u}{x}\Delta_1(x, K)$$

and

$$v' = \frac{1}{x} B(x, K)\left(\frac{1}{2}x_2\right)\left(u + \frac{1}{2}x_2\right) - \frac{v}{x}\Delta_1(x, K) + \frac{x_3}{x}(\Delta_1(x, K) - \Delta_2(x, K))$$

$$\leq \frac{1}{x} B(x, K)\left(u + \frac{1}{2}x_2\right)v - \frac{v}{x}\Delta_1(x, K) + \frac{x_3}{x}(\Delta_1(x, K) - \Delta_2(x, K)).$$

This yields that

$$\frac{v'}{v} - \frac{u'}{u} \leq \frac{x_3}{vx}(\Delta_1(x, K) - \Delta_2(x, K)).$$

Integrating both sides produces

$$\ln\frac{v}{v_0} - \ln\frac{u}{u_0} \leq \int_0^t \frac{x_3}{vx}(\Delta_1(x, K) - \Delta_2(x, K))\,ds,$$

and hence

$$v \leq cu \exp\left(\int_0^t \frac{x_3}{vx}(\Delta_1(x, K) - \Delta_2(x, K))\,ds\right).$$

Since $(x_3/vx)(\Delta_1(x, K) - \Delta_2(x, K)) \leq 0$, $M = \lim_{t\to\infty}\int_0^t (x_3/vx)(\Delta_1(x, K) - \Delta_2(x, K))\,dt$ exists. If $M = -\infty$, then $\lim_{t\to\infty} v = 0$. If $M > -\infty$, then $\lim_{t\to\infty}(x_3/vx)(\Delta_1(x, K) - \Delta_2(x, K)) = 0$. By (H2), $(\Delta_1(x, K) - \Delta_2(x, K))/x \geq \delta > 0$, so it must be that $\lim_{t\to\infty}(x_3(t)/v(t)) = 0$. Since v is bounded, $\liminf_{t\to\infty} x_3(t) = 0$, and Lemma 3.2 implies that $\liminf_{t\to\infty} x_2(t) = 0$. If $\limsup_{t\to\infty} v(t) = \varepsilon_0 > 0$, choose sequences t_n and t'_n, $t_n \to \infty$ and $t'_n \to \infty$ as $n \to \infty$, such that $v(t_n) = \frac{1}{3}\varepsilon_0$, $v(t'_n) = \frac{2}{3}\varepsilon_0$, and

$$\tfrac{1}{3}\varepsilon_0 < v(t) < \tfrac{2}{3}\varepsilon_0$$

for $t_n < t < t'_n$. Let $p_n = \min\{x_3(t) \mid t_n \le t \le t'_n\}$. If there were a subsequence n_k such that

$$p_{n_k} \to 0 \quad \text{as } n_k \to \infty,$$

then there would be a sequence $\{s_n\}$, $s_n \to \infty$ as $n \to \infty$, such that $x_3(s_n) \to 0$ as $n \to \infty$. However, $v(s_n) \ge \frac{1}{3}\varepsilon_0$, contradicting Lemma 3.2. Therefore, there are $\delta_0 > 0$ and $N > 0$ such that $p_n \ge \delta_0 > 0$ for $n \ge N$. Now $v'(t) \le \beta$ for some $\beta > 0$ and for all t, so

$$t_n - t'_n > \frac{\varepsilon_0}{3\beta}.$$

However, if this were the case, then

$$\int_{t_n}^{t'_n} \frac{x_3(s)}{v(s)}\, ds \ge \frac{\delta_0}{\varepsilon_0}\frac{\varepsilon_0}{3\beta} = \frac{\delta_0}{3\beta},$$

which contradicts the convergence of

$$\int_0^\infty \frac{x_3(s)}{v(s)}\, ds.$$

Therefore $\lim_{t\to\infty} v(t) = 0$.

Before turning to the proof of Theorem 3.1, two lemmas are stated; both are computational in nature and the proofs are omitted.

LEMMA 3.3. *If $x_3(0) \le x_1(0)$ then $x_3(t) \le x_1(t)$ for all $t \ge 0$.*

LEMMA 3.4. *If $x_1(0) > 0$, $x_3(0) < x_1(0)$, and if $x_3(0)/x_2(0) \le x_2(0)/x_1(0)$, then $x_3(t)/x_2(t) \le x_2(t)/x_1(t)$ for $t > 0$.*

Proof of Theorem 3.1. Let $u = x_1 + \frac{1}{2}x_2$, $v = x_3 + \frac{1}{2}x_2$. Then

$$u' = \frac{u(x_1 + (\alpha/2)x_2)}{x} B(x, K) - \frac{u}{x}\Delta_1(x, K)$$

$$+ \frac{1}{x^2} B(x, K)\left(\frac{1}{4}x_2 x_1 + \frac{\alpha}{4}x_2 x_1 + \frac{\alpha}{4}x_2^2\right)$$

and

$$v' = \frac{1}{x} B(x, K)\left[\frac{1}{4}x_2 x_1 + \frac{\alpha}{4}x_2 x_1 + \frac{\alpha}{4}x_2^2 + \frac{\alpha}{4}x_2^2\right]$$

$$- \frac{v}{x}\Delta_1(x, K) + \frac{x_3}{vx}(\Delta_1(x, K) - \Delta_2(x, K)).$$

Thus

$$\frac{v'}{v} - \frac{u'}{u} = \frac{B(x,K)}{xv}\left[\frac{1}{4}x_2x_1 + \frac{\alpha}{4}x_2x_1 + \frac{\alpha}{2}x_2^2\right]$$

$$- \frac{B(x,K)}{x}\left[x_1 + \frac{\alpha}{2}x_2\right] + \frac{x_3}{vx}(\Delta_1(x,K) - \Delta_2(x,K))$$

or

(3.4) $\quad v = cu \exp\left\{\int_0^t \left[\frac{B(x,K)}{x}\left(\frac{1}{4}\frac{x_2x_1}{v} + \frac{\alpha}{4}\frac{x_2x_1}{v} + \frac{\alpha}{2}\frac{x_2^2}{v} - x_1 - \frac{\alpha}{2}x_2\right)\right.\right.$

$$\left.\left. + \frac{x_3}{vx}(\Delta_1(x,K) - \Delta_2(x,K))\right]ds\right\}.$$

Assume that $\lim_{t\to\infty} v(t) = 0$, which implies that $\lim_{t\to\infty} x_1(t) = \lim_{t\to\infty} u(t) = L$. Then it must be the case that

$$\lim_{t\to\infty}\int_0^t\left[\frac{B(x,K)}{x}\left(\frac{1}{4}\frac{x_2x_1}{v} + \frac{\alpha}{4}\frac{x_2x_1}{v} + \frac{\alpha}{2}\frac{x_2^2}{v} - x_1 - \frac{\alpha}{2}x_2\right)\right.$$

$$\left. + \frac{x_3}{vx}(\Delta_1(x,K) - \Delta_2(x,K))\right]ds = -\infty.$$

Since

$$\frac{x_3}{v} = \frac{x_3}{x_3 + \frac{1}{2}x_2} = \frac{1}{1 + \frac{1}{2}x_2/x_3} \quad \text{and} \quad \frac{x_3}{x_2} < \frac{x_2}{x_1}$$

(Lemma 3.4), then $\lim_{t\to\infty}(x_3/x_2) = 0$, and hence $\lim_{t\to\infty}(x_3/v) = 0$. Similarly, $\lim_{t\to\infty}(x_2/v) = 2$. Now, taking the limit of the integrand gives

$$\lim_{t\to\infty}\frac{B(x,K)}{x}\left(\frac{1}{4}\frac{x_2x_1}{v} + \frac{\alpha}{4}\frac{x_2x_1}{v} + \frac{\alpha}{2}\frac{x_2^2}{v} - x_1 - \frac{\alpha}{2}x_2\right) + \frac{x_3}{vx}(\Delta_1(x,K) - \Delta_2(x,K))$$

$$= \frac{B(x,L)}{L}\left(\frac{1}{2}L + \frac{\alpha}{2}L - L\right) > 0,$$

which is a contradiction. Therefore $v(t)$ cannot approach zero as $t \to \infty$.

Now assume that there is a sequence t_n, $t_n \to \infty$ as $n \to \infty$, such that $v(t_n) \to 0$ as $n \to \infty$. Define

$$F(t) = \int_0^t\left[\frac{B(x,K)}{x}\left(\frac{\alpha}{4}\frac{x_2x_1}{v} + \frac{\alpha}{4}\frac{x_2x_1}{v} + \frac{\alpha}{2}\frac{x_2^2}{v} - x_1 - \frac{\alpha}{2}x_2\right)\right.$$

$$\left. + \frac{x_3}{vx}(\Delta_1(x,K) - \Delta_2(x,K))\right]ds.$$

If F is eventually monotone, then $\lim_{t\to\infty} F(t) = -\infty$ and hence, using (3.4), $\lim_{t\to\infty} v(t) = 0$ which we have already shown leads to a contradiction. Therefore the t_n can be chosen such that $F(t_n)$ is a local minimum and hence $F'(t_n) = 0$. But as before,

$$\lim_{n\to\infty} F'(t_n) = \frac{B(x,K)}{L}\left(\frac{1}{2}L + \frac{\alpha}{2}L - L\right) > 0,$$

which is a contradiction. Therefore, $\lim\inf_{t\to\infty} v(t) > 0$, which proves the theorem.

Taken together, these two theorems state that, without any added advantage, the gene for cystic fibrosis would die out of the population but, with an advantage in the number of offspring, it does not.

To simulate the disease in the United States, the equations (3.3) are made more specific, taking into account the more general logistic equation introduced earlier to describe the evolution of the population in the United States. The model used for the simulation was

(3.5)
$$x_1' = \frac{\beta}{x}\left(x_1 + \frac{\alpha}{2}x_2\right)\left(x_1 + \frac{1}{2}x_2\right) - \frac{\beta x_1 x}{K},$$

$$x_2' = \frac{\beta}{x}\left(\frac{1}{2}x_2 x_1 + \frac{\alpha}{2}x_2 x_1 + \frac{\alpha}{2}x_2\right) - \frac{\beta x_2 x}{K},$$

$$x_3' = \frac{\beta}{x}\left(\frac{\alpha}{4}x_2^2\right) - \frac{\beta x_3 x}{K} - d x_3,$$

$$K' = \gamma K - \gamma \frac{K^2}{L},$$

$$x_1(0) = x_{10}, \quad x_2(0) = x_{20}, \quad x_3(0) = x_{30}, \quad K(0) = K_0, \quad x = x_1 + x_2 + x_3.$$

This corresponds to $B(x,K) = \beta$, $\Delta_1(x,K) = \beta x^2/K$, and $\Delta_2(x) = \Delta_1(x) - d$.

Parameter values were taken as follows: $\beta = .0322$, $x_{10} = 2.759$ (million), $K_0 = 184$, $L = 350$, $\gamma = .008$, $d = 4$, $x_{20} = .001$, $x_{30} = 0$. As mentioned above, control families in the observations in [2] had an average of 4.03 offspring compared to 4.67 for grandparents of cystic fibrosis patients. If the selective advantage exists only for the female heterozygote, then half of the study group must have had a normal birth rate while the other half had a higher birth rate. In terms of the model, $\frac{1}{2}(4.03 + \alpha 4.03) = 4.67$ and so $\alpha = 1.318$. Simulations with this value of α indicate that the occurrence of the cystic fibrosis gene is increasing at present. An initial value of $x_{20} = .05$ was chosen so that an incidence of cystic fibrosis births of 1 in 2000 births occurs in the range from 1950 to 1970. Simulations for the future suggest that with the above parameters the incidence will level out at around 14 cystic fibrosis births out of every 1000 births. See Fig. 3.2.

If, on the other hand, it is assumed that the incidence of cystic fibrosis has already stabilized at 1 out of every 2000 births, the simulation shows that an assumption of $\alpha = 1.0468$ is appropriate and that the number of heterozygotes in 1970 was approximately .126 million. The difference between this value of α and the observed value is substantial.

One caution is appropriate in interpreting the simulations—it was assumed that α was constant. Modern birth control methods, improved genetic testing, and appro-

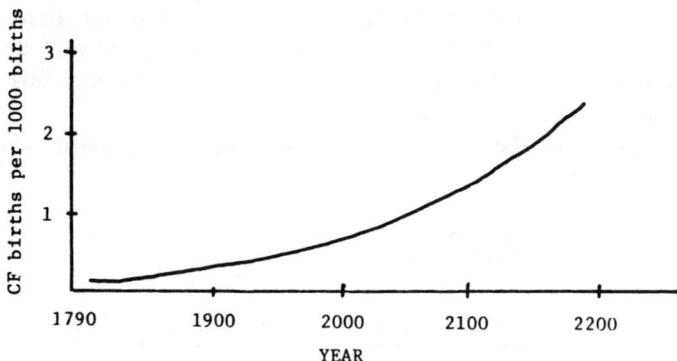

FIG. 3.2. *Simulation of the evolution of cystic fibrosis in the United States, from Beck* [5]. [Copyright 1982, Elsevier Science Publishing Co. Used by permission.]

priate counseling serve to lower α and reduce the projections. Finally, one should note that mutation has been ignored and that there may be other ways to introduce an advantage for the heterozygote than the one used in this model.

The effects of a disease on evolution have been explored further in Beck, Keener and Ricciardi [6], [7].

4. A parental selection problem. Suppose that in the model of a one-locus, two-allele system there is no death rate and that the overall intrinsic growth rate is constant. The equations (1.4) take the form

(4.1)
$$x_1' = \frac{(x_1 + \tfrac{1}{2} x_2)^2}{x} = \frac{(x_1)^2 + (x_1)(x_2) + \tfrac{1}{4} (x_2)^2}{x},$$

$$x_2' = \frac{2(x_1 + \tfrac{1}{2} x_2)(x_3 + \tfrac{1}{2} x_2)}{x}$$

$$= 2\left[\frac{(x_1)(x_3) + \tfrac{1}{2} (x_2)(x_3) + \tfrac{1}{2} (x_2)(x_1) + \tfrac{1}{4} (x_2)^2}{x}\right],$$

$$x_3' = \frac{(x_3 + \tfrac{1}{2} x_2)^2}{x} = \frac{(x_3)^2 + (x_3)(x_2) + \tfrac{1}{4} (x_2)^2}{x},$$

$$x_i(0) \geq 0, \quad i = 1, 2, 3, \quad x_1(0) + x_2(0) + x_3(0) > 0.$$

x_i/x represents the proportion of genotype i in the population. (4.1) corresponds to an exponentially growing population since adding the equations yields

$$x' = x.$$

The model to be studied in this section assumes that each genotype has equal survivability (in contrast to the cystic fibrosis case) and that the number of offspring is a function of the genotype of the parents. (This aspect was partially present in §3 where the heterozygote mother had more children.) The interest in such a problem is to study the eventual behavior of solutions of the system of equations governing the growth process—the eventual distribution of the genes. Steady states which represent a combination of different genotypes are called polymorphisms, and there is interest in determining the possible stable polymorphisms.

If one supposes that there are fertility differences in the matings, the model becomes (see §1, (1.5))

$$x_1' = \frac{x_1^2 B_{11} + x_1 x_2 B_{12} + \tfrac{1}{4} x_2^2 B_{22}}{x},$$

$$x_2' = 2\left[\frac{x_1 x_3 B_{13} + \tfrac{1}{2} x_2 x_3 B_{23} + \tfrac{1}{2} x_1 x_2 B_{12} + \tfrac{1}{4} x_2^2 B_{22}}{x}\right],$$

$$x_3' = \frac{x_3^2 B_{33} + x_2 x_3 B_{23} + \tfrac{1}{4} x_2^2 B_{22}}{x}.$$

The term B_{ij} denotes the combined parental fertility of the ith and jth genotype. Ideally B_{ij} might be a density dependent function, but the analysis, even when it is constant for each i, j—an assumption that we make here—is nontrivial. By a change of time scale we can assume $0 \leq B_{ij} \leq 1$.

Hadeler and Liberman [28] have studied such fertility models in the discrete case and have analyzed one example (numerically) in the continuous model. In keeping with the spirit of these lectures, it appears desirable to have differential equation models for the genetics to facilitate the combination with continuous ecological models. This section follows Butler, Freedman and Waltman [11] and Hadeler and Glass [27]. With the assumption of considerable symmetry it will be shown that the limiting ratio $x_i(t)/x(t)$ exists and the possible outcomes will be determined as a function of the fertility parameters.

As noted above, particular interest is focused on the possible polymorphic states; so as not to prejudice the outcome in favor of a homozygote, it is reasonable to assume that $B_{11} = B_{33} = \alpha$ and $B_{12} = B_{23} = \gamma$. This might be called the symmetric case. Here the further assumption $B_{13} = \alpha$ will also be made—a totally symmetric case. Writing $B_{22} = \beta$, (4.1) becomes

(4.2)
$$x_1' = \frac{1}{x}\left[\alpha x_1^2 + \gamma x_1 x_2 + \frac{1}{4}\beta x_2^2\right],$$

$$x_2' = \frac{2}{x}\left[\alpha x_1 x_3 + \frac{\gamma}{2} x_2 x_3 + \frac{\gamma}{2} x_1 x_2 + \frac{1}{4}\beta x_2^2\right],$$

$$x_3' = \frac{1}{x}\left[\alpha x_3^2 + \gamma x_3 x_2 + \frac{1}{4}\beta x_2^2\right].$$

Since solutions to (4.2) are unbounded and since it is the limiting ratios that are relevant, it is convenient to change variables to $y_i = x_i/x$—to the frequencies of the genotypes. The corresponding differential equations become

(4.3)
$$\begin{aligned}
y_1' &= \alpha(y_1^2 - 2y_1^2 y_3 - y_1 y_3^2 - y_1^3) \\
&\quad + \beta(\tfrac{1}{4} y_2^2 - y_1 y_2^2) + \gamma(y_1 y_2 - 2y_1^2 y_2 - 2y_1 y_2 y_3), \\
y_2' &= \alpha(-y_1^2 y_2 + 2y_1 y_3 - 2y_1 y_2 y_3 - y_2 y_3^2) \\
&\quad + \beta(\tfrac{1}{2} y_2^2 - y_2^3) + \gamma(y_1 y_2 - 2y_1 y_2^2 + y_2 y_3 - 2y_2^2 y_3), \\
y_3' &= \alpha(-y_1^2 y_3 - 2y_1 y_3^2 + y_3^2 - y_3^3) \\
&\quad + \beta(\tfrac{1}{4} y_2^2 - y_2^2 y_3) + \gamma(-2y_1 y_2 y_3 + y_2 y_3 - 2y_2 y_3^2),
\end{aligned}$$

$$y_i(0) \geq 0, \quad y_1(t) + y_2(t) + y_3(t) = 1.$$

The constraint that the variables sum to one means that we are effectively dealing with a two-dimensional system. We note that even if (4.3) is regarded as a proper three-dimensional system, disregarding the constraint $y_1 + y_2 + y_3 = 1$, provided that $\beta \neq 0$, the planar region $y_1 + y_2 + y_3 = 1$ will be globally asymptotically stable with respect to the positive octant, which is invariant with respect to (4.3) (Lemma 4.1). This is important for numerical calculations since the initial constraint $y_1(0) + y_2(0) + y_3(0) = 1$ may not be achieved in the machine's arithmetic.

THEOREM 4.1. *For each set of allowable initial conditions, the limit of the frequencies defined by system* (4.2) *exists, i.e.,*

$$\lim_{t \to \infty} \frac{x_i(t)}{x(t)} = K_i \geq 0, \quad i = 1, 2, 3.$$

One objective is to determine and classify the equilibria of (4.3) according to their stability as a function of the fertility parameters. Those equilibria which are interior to the positive octant correspond to the possible polymorphic states and are called interior equilibria. Those which lie on the boundary correspond to states with one or more genotypes missing and are called boundary equilibria. The interior equilibria are further divided into those with coordinates of the form (y_1^*, y_2^*, y_1^*), said to be symmetric, and the remainder, said to be asymmetric.

THEOREM 4.2. *If $\beta \neq 0$, $\alpha \neq 0$, the only symmetric interior equilibrium is $(\tfrac{1}{4}, \tfrac{1}{2}, \tfrac{1}{4})$. This equilibrium is asymptotically stable if $\beta > \alpha$ and unstable if $\beta < \alpha$.*

THEOREM 4.3. *Let $A = (\beta - \gamma)/(\alpha - \gamma)$. If $A > 0$ and if $A^2 - \beta/\alpha > 0$, asymmetrical, interior equilibria exist. If $\beta > \alpha > \gamma$, the asymmetric interior equilibria are unstable; if $\alpha < \gamma$, they are asymptotically stable.*

Theorem 4.3 is a complete result since $\alpha < \gamma$ will force $\beta < \alpha$ if $A^2 > \beta/\alpha$.

The equilibria on the boundary will be located and analyzed in the course of proving the above theorems, so we defer discussion of them until then. As noted above, in view of Theorem 4.1, the asymptotically stable interior equilibria represent the only possible polymorphic states of all three genotypes. A symmetric polymorphism means that the eventual population is unprejudiced by the parameter choices, while an asymmetric one biases the relative population toward one genotype.

As noted above, analysis of the system in the region T given by $y_1 + y_2 + y_3 = 1$, $y_i \geq 0$, $i = 1, 2, 3$, is sufficient to establish dynamical behavior for the whole of the positive octant. This is the content of the following lemma.

LEMMA 4.1. *Let $\beta \neq 0$. The region T is globally asymptotically stable with respect to the positive octant for the system* (4.3).

Proof. The proof follows by adding the equations in (4.3) and simplifying to obtain

$$(y_1 + y_2 + y_3)' = (1 - y_1 - y_2 - y_3)$$
$$\cdot (\alpha y_1^2 + \beta y_2^2 + \alpha y_3^2 + 2\gamma y_2(y_1 + y_3) + 2\alpha y_1 y_3).$$

Since $\beta > 0$ and $\alpha, \gamma \geq 0$, the quadratic factor in the above equation is positive in the positive octant. The stability properties of T follow.

If $\beta = 0$, the plane $y_1 + y_2 + y_3 = 1$ remains locally stable and its basin of attraction is all of the positive octant except the line $y_1 = y_3 = 0$.

Preliminary to finding the equilibria of (4.3) we utilize the fact that $y_1 + y_2 + y_3 = 1$ to reduce system (4.3) to a two-dimensional system in y_1 and y_3. Substituting $y_2 = 1 - y_1 - y_3$ gives

$$y_1' = \tfrac{1}{4}\beta + (-\tfrac{3}{2}\beta + \gamma)y_1 + (\alpha + \tfrac{9}{4}\beta - 3\gamma)y_1^2 + (-\alpha - \beta + 2\gamma)y_1^3$$
$$+ (\tfrac{5}{2}\beta - 3\gamma)y_1 y_3 + (-2\alpha - 2\beta + 4\gamma)y_1^2 y_3 + (-\alpha - \beta + 2\gamma)y_1 y_3^2$$
$$- \tfrac{1}{2}\beta y_3 + \tfrac{1}{4}\beta y_3^2,$$

(4.4)

$$y_3' = \tfrac{1}{4}\beta + (-\tfrac{3}{2}\beta + \gamma)y_3 + (\alpha + \tfrac{9}{4}\beta - 3\gamma)y_3^2 + (-\alpha - \beta + 2\gamma)y_3^3$$
$$+ (\tfrac{5}{2}\beta - 3\gamma)y_1 y_3 + (-2\alpha - 2\beta + 4\gamma)y_1 y_3^2 + (-\alpha - \beta + 2\gamma)y_1^2 y_3$$
$$- \tfrac{1}{2}\beta y_1 + \tfrac{1}{4}\beta y_1^2,$$

$$y_1(0) \geq 0, \quad y_3(0) \geq 0, \quad 0 \leq y_1(0) + y_3(0) \leq 1.$$

An equilibrium of system (4.4) of the form (y_1^*, y_3^*) transforms into an equilibrium of system (4.3) of the form $(y_1^*, 1 - y_1^* - y_3^*, y_3^*)$.

Making the further change of variables $u = y_1 + y_3$, $v = y_1 - y_3$ transforms the system (4.4) to

$$u' = \tfrac{1}{2}\beta + (-2\beta + \gamma)u + (\tfrac{1}{2}\alpha + \tfrac{5}{2}\beta - 3\gamma)u^2$$
(4.5)
$$+ (-\alpha - \beta + 2\gamma)u^3 + \tfrac{1}{2}\alpha v^2,$$

$$v' = v[(-\beta + \gamma) + (\alpha + 2\beta - 3\gamma)u + (-\alpha - \beta + 2\gamma)u^2],$$

$$0 \leq u(0) \leq 1, \quad -1 \leq v(0) \leq 1.$$

Equilibria of the form (u^*, v^*) of system (4.5) transform into equilibria of the form $(\tfrac{1}{2}(u^* + v^*), \tfrac{1}{2}(u^* - v^*))$ of system (4.4).

To find the equilibria of system (4.5), solve the system of algebraic equations

$$f(u, v) = \tfrac{1}{2}\beta + (-2\beta + \gamma)u + (\tfrac{1}{2}\alpha + \tfrac{5}{2}\beta - 3\gamma)u^2$$
(4.6)
$$+ (-\alpha - \beta + 2\gamma)u^3 + \tfrac{1}{2}\alpha v^2 = 0,$$

$$g(u, v) = v[(-\beta + \gamma) + (\alpha + 2\beta - 3\gamma)u + (-\alpha - \beta + 2\gamma)u^2] = 0.$$

Clearly, the second of these is satisfied if $v = 0$ (giving rise to the symmetric equilibria). Now

(4.7) $\qquad f(u, 0) = (u - \tfrac{1}{2})[(-\alpha - \beta + 2\gamma) u^2 + (2\beta - 2\gamma) u - \beta].$

Clearly one equilibrium is E_1'': $(\tfrac{1}{2}, 0)$.

E_1'' is the only equilibrium of the form $(u^*, 0)$, for if $(u^*, 0)$ were an equilibrium of (4.5), then $f(u^*, 0) = 0$ and $0 \leq u^* \leq 1$. If $2\gamma = \alpha + \beta$, then $u^* = \beta/(2\beta - 2\gamma) = \beta/(\beta - \alpha)$, $u^* > 1$ if $\beta - \alpha > 0$, or $u^* < 0$ if $\beta - \alpha < 0$. Thus there is no feasible equilibrium in this case.

Suppose, then, that $2\gamma \neq \alpha + \beta$. Then from (4.7),

(4.8) $\qquad u^* = \dfrac{\gamma - \beta \pm (\gamma^2 - \alpha\beta)^{1/2}}{2\gamma - \alpha - \beta}.$

If $\gamma^2 - \alpha\beta < 0$, u^* is not real. If $\gamma^2 - \alpha\beta = 0$,

$$u^* = \frac{\gamma - \beta}{2\gamma - \alpha - \beta} = \frac{\gamma - \beta}{-(\sqrt{\alpha} - \sqrt{\beta})^2}.$$

Hence, to have $u^* > 0$, one must have $\alpha < \beta$. However, to have $u^* \leq 1$ with $2\gamma - \alpha - \beta < 0$, it must be the case that $\gamma - \beta \geq 2\gamma - \alpha - \beta$ or $\gamma \leq \alpha$. Hence $\gamma^2 < \alpha\beta$, a contradiction.

Finally consider the case where $\gamma^2 - \alpha\beta > 0$. Suppose $2\gamma - \alpha - \beta > 0$. (The analysis in the case $2\gamma - \alpha - \beta < 0$ proceeds in a similar manner.) Then to have

$$u^* = \frac{\gamma - \beta - (\gamma^2 - \alpha\beta)^{1/2}}{2\gamma - \alpha - \beta} > 0$$

it must be the case that $\gamma - \beta > (\gamma^2 - \alpha\beta)^{1/2} > 0$. Upon squaring and simplifying, we get $-2\gamma\beta + \beta^2 > -\alpha\beta$, or $2\gamma - \alpha - \beta < 0$, a contradiction.
If

$$u^* = \frac{\gamma - \beta + (\gamma^2 - \alpha\beta)^{1/2}}{2\gamma - \alpha - \beta},$$

then $u^* \leq 1$ implies that $(\gamma^2 - \alpha\beta)^{1/2} \leq \gamma - \alpha$. Simplification yields $2\gamma - \alpha - \beta \leq 0$, a contradiction.

This shows that E_1'' is in the only nondegenerate equilibrium with $v^* = 0$. There is also a degenerate case when $\beta = 0$. Then E_0'': $(0, 0)$ is an equilibrium (the other value for u^*, in this case $2\gamma/(2\gamma - \alpha)$, again falls outside the range $0 \leq u^* \leq 1$).

In order to study the stability of these equilibria, compute the variational matrix of (4.5)

(4.9)

$$V(u, v) = \begin{bmatrix} (-2\beta + \gamma) + (\alpha + 5\beta - 6\gamma) u \\ + 3(-\alpha - \beta + 2\gamma) u^2 & \alpha v \\ (\alpha + 2\beta - 3\gamma) v & (-\beta + \gamma) + (\alpha + 2\beta - 3\gamma) u \\ + 2(-\alpha - \beta + 2\gamma) uv & + (-\alpha\beta + 2\gamma) u^2 \end{bmatrix}.$$

Let $V(0, 0) = V_0 (\beta = 0)$ and $V(\frac{1}{2}, 0) = V_1$. Then

$$V_0 = \begin{bmatrix} \gamma & 0 \\ 0 & \gamma \end{bmatrix} \quad \text{and} \quad V_1 = \begin{bmatrix} \frac{1}{4}(-\alpha - \beta - 2\gamma) & 0 \\ 0 & \frac{1}{4}(\alpha - \beta) \end{bmatrix}.$$

This shows that E_0'' is unstable and E_1'' is asymptotically stable if $\alpha < \beta$, and unstable if $\alpha > \beta$.

Note that E_0'' translates into E_0': $(0, 0)$ in the (y_1, y_3)-plane, and into E_0: $(0, 1, 0)$ in (y_1, y_2, y_3)-space, E_1'' translates into E_1': $(\frac{1}{4}, \frac{1}{4})$ in the (y_1, y_3)-plane, and into E_1: $(\frac{1}{4}, \frac{1}{2}, \frac{1}{4})$ in (y_1, y_2, y_3)-space. This completes the proof of Theorem 4.2.

We now return to the algebraic equations (4.6) and seek equilibria (u, v) with $v \neq 0$. Then $v^{-1} g(u, v) = 0$. This has roots of the form

$$\mu^* = 1, \qquad u^* = \frac{\beta - \gamma}{\alpha + \beta - 2\gamma}.$$

If $u^* = 1$, then solving $f(1, v) = 0$, we have $v^* = 1$ or -1. This leads to two equilibria of the form E_2'': $(1, 1)$ and E_3'': $(1, -1)$, which transform into E_2: $(1, 0, 0)$ and E_3: $(0, 0, 1)$ in (y_1, y_2, y_3)-space.

Let

(4.10) $$u = \frac{\beta - \gamma}{\alpha + \beta - 2\gamma}.$$

Then for $u^* = \mu$ to be such that $0 < u^* \leq 1$, it follows that either $\alpha > \gamma, \beta > \gamma$, or $\alpha < \gamma, \beta < \gamma$.

Suppose that one of the above holds. Then v^* is given by

$$\tfrac{1}{2} \alpha v^{*2} = (\mu - \tfrac{1}{2})[(\alpha + \beta - 2\gamma) \mu^2 - 2(\beta - \gamma) \mu + \beta].$$

This reduces, upon substitution from (4.10) and simplification, to

(4.11) $$\alpha v^{*2} = \frac{(\beta - \alpha)(\alpha \beta - \gamma^2)}{(\alpha + \beta - 2\gamma)^2}.$$

The right-hand side of (4.11) must be positive. If $\alpha > \gamma$ and $\beta > \gamma$, then $\beta > \alpha$ must hold. Similarly, if $\alpha < \gamma$ and $\beta < \gamma$, then $\beta < \alpha$. Hence such equilibria exist if and only if either $\beta > \alpha > \gamma$ or $\beta < \alpha < \gamma$. Suppose one of these inequalities holds. If v is defined by

$$v = \left(\frac{(\beta - \alpha)(\alpha \beta - \gamma^2)}{\alpha} \right)^{1/2} \frac{1}{|\alpha + \beta - 2\gamma|},$$

then E_4'': (μ, v) and E_5'': $(\mu, -v)$ are equilibria. These transform to E_4: $(\frac{1}{2}(\mu + v), 1 - \mu, \frac{1}{2}(\mu - v))$ and E_5: $(\frac{1}{2}(\mu - v), 1 - \mu, \frac{1}{2}(\mu + v))$.

We now examine the stability of these asymmetrical equilibria. First consider E_2'' and E_3''; by substituting $u = 1$ and $v = \pm 1$ into (4.9), one obtains

$$V_2 = \begin{bmatrix} -2\alpha + \gamma & \alpha \\ -\alpha + \gamma & 0 \end{bmatrix}, \quad V_3 = \begin{bmatrix} -2\alpha + \gamma & -\alpha \\ \alpha - \gamma & 0 \end{bmatrix},$$

respectively. In both cases the eigenvalues are $-\alpha$ and $\gamma - \alpha$. Hence E_2 and E_3 are asymptotically stable if $\gamma < \alpha$ and are unstable if $\gamma > \alpha$.

If E_4'' and E_5'' exist, then substituting $u = \mu$ and $v = \pm v$ into (4.9) yields

$$V_4 = \begin{bmatrix} \dfrac{\alpha\beta - \gamma^2}{-\alpha - \beta + 2\gamma} & \alpha v \\ (\alpha - \gamma)v & 0 \end{bmatrix}, \quad V_5 = \begin{bmatrix} \dfrac{\alpha\beta - \gamma^2}{-\alpha - \beta + 2\gamma} & -\alpha v \\ -(\alpha - \gamma)v & 0 \end{bmatrix}.$$

The eigenvalues in both cases are

$$(4.12) \quad \lambda = \frac{1}{2}\left(\frac{\alpha\beta - \gamma^2}{-\alpha - \beta + 2\gamma}\right) \pm \frac{1}{2}\left[\left(\frac{\alpha\beta - \gamma^2}{-\alpha - \beta + 2\gamma}\right)^2 + 4\alpha(\alpha - \gamma)v^2\right]^{1/2}.$$

Clearly then, if $\beta < \alpha < \gamma$ or $\beta > \alpha > \gamma$ (the conditions for these equilibria to exist), $(\alpha\beta - \gamma^2)/(-\alpha - \beta + 2\gamma) < 0$. Further, if $\beta < \alpha < \gamma$, then $4\alpha(\alpha - \gamma)v^2 < 0$ and E_4 and E_5 are asymptotically stable. If $\beta > \alpha > \gamma$, then $4\alpha(\alpha - \gamma)v^2 > 0$ and E_4 and E_5 are saddle points.

This proves Theorem 4.3. The above stability results are summarized in Table 4.1 (except for E_0 which only occurs in the degenerate case of $\beta = 0$), reproduced from [11].

To make the above local argument global, there remains to show that the omega limit sets of all trajectories are critical points. Since the system being analyzed is two-dimensional, the Poincaré–Bendixson theorem allows the conclusion that the omega limit set is a critical point, a periodic solution, or critical points and trajectories "joining" them. "Joining" means critical points and orbits which have the critical points as alpha or omega limit points in such a way that the union is a connected set.

Clearly, for this to happen the critical points involved must be saddle points (repellers cannot be in the omega limit set and attractors must be all of the omega limit set). The line $y_1 = y_3$ is invariant under the solution flow ($v = 0$ in the transformed system). This has two consequences: (i) the region T (of Lemma 4.1) is divided into two invariant halves—no trajectory may cross $y_1 = y_3$—and (ii) if E is a saddle point, the stable manifold must lie on $y_1 = y_3$. Consulting Table 4.1, one sees that there is only one case, case IV, where two saddle points exist in either of the closed halves of T created by the dividing plane $y_1 = y_3$. However, since the stable manifold of E_1 lies in the $y_1 = y_3$ plane, E_1 cannot be the omega limit set of any trajectory for which E_2 or E_3 is the alpha limit set. Thus to establish a global theorem it is necessary only to eliminate limit cycles. Although a proof of this was

TABLE 4.1

	$E_1(E_1)$	$E_2, E_3(E_2', E_3')$	$E_4, E_5(E_4', E_5')$
(I) $\alpha > \beta, \gamma$	Unstable	Stable	Nonexistent
(II) $\alpha < \beta, \gamma$	Stable	Unstable	Nonexistent
(III) $\beta > \alpha > \gamma$	Stable	Stable	Unstable (saddle point)
(IV) $\beta < \alpha < \gamma$	Unstable	Unstable	Stable

given in [11], a shorter and more general proof appears in Hadeler and Glass [27]. If a transformation is made in the system (4.2) to new variables ξ and η by

$$\xi = \frac{x_1}{x_2} \quad \text{and} \quad \eta = \frac{x_3}{x_2},$$

then the region $x_2 > 0$ is mapped onto the interior of the positive quadrant in the (ξ, η)-plane. Let $(x_1(t), x_2(t), x_3(t))$ be a given nontrivial, periodic solution of (4.2); necessarily $\inf_t x_2(t) > 0$. Then (4.2) may be written

(4.13)
$$x_1' = \frac{x_2^2}{x}\left[\alpha\xi^2 + \gamma\xi + \frac{1}{4}\beta\right],$$

$$x_2' = \frac{2x_2^2}{x}\left[\alpha\xi\eta + \frac{\gamma}{2}\eta + \frac{\gamma}{2}\xi + \frac{1}{4}\beta\right],$$

$$x_3' = \frac{x_2^2}{x}\left[\alpha\eta^2 + \gamma\eta + \frac{1}{4}\beta\right].$$

The equations for ξ and η become

$$\xi' = \frac{x_2 x_1' - x_1 x_2'}{x_2^2} = \frac{x_1' - \xi x_2'}{x_2}$$

$$= \frac{x_2}{x}\left[\alpha\xi^2 + \gamma\xi + \frac{1}{4}\beta - 2\alpha\xi^2\eta - \gamma\xi\eta - \gamma\xi^2 - \xi\frac{\beta}{2}\right]$$

(4.14)
$$= \frac{x_2}{x}\left[\xi^2(\alpha - \gamma) + \xi\left(\gamma - \frac{\beta}{2}\right) - 2\alpha\xi^2\eta - \gamma\xi\eta + \frac{1}{4}\beta\right],$$

$$\eta' = \frac{x_2 x_3' - x_3 x_2'}{x_2^2} = \frac{x_3' - \eta x_2'}{x_2}$$

$$= \frac{x_2}{x}\left[\eta^2(\alpha - \gamma) + \eta\left(\gamma - \frac{\beta}{2}\right) - 2\alpha\xi\eta^2 + \frac{1}{4}\beta\right].$$

Making a (solution-dependent) time scale change $\tau = \int_0^t (x_2(s)/x(s))\,ds$ changes (4.14) to

(4.15)
$$\xi' = (\alpha - \gamma)\xi^2 + \left(\gamma - \frac{\beta}{2}\right)\xi - 2\alpha\xi^2\eta - \gamma\xi\eta + \frac{1}{4}\beta,$$

$$\eta' = (\alpha - \gamma)\eta^2 + \left(\gamma - \frac{\beta}{2}\right)\eta - 2\alpha\xi\eta^2 - \gamma\xi\eta + \frac{1}{4}\beta, \quad ' = \frac{d}{d\tau}.$$

If one writes

(4.16)
$$\xi' = f(\xi, \eta), \quad \eta' = g(\xi, \eta)$$

then

(4.17)
$$f_\eta = -2\alpha\xi^2 - \gamma\xi < 0, \quad g_\xi = -2\alpha\eta^2 - \gamma\eta < 0.$$

Hence the inequality of Kamke (in §2) would apply if we replaced τ by $-\tau$.

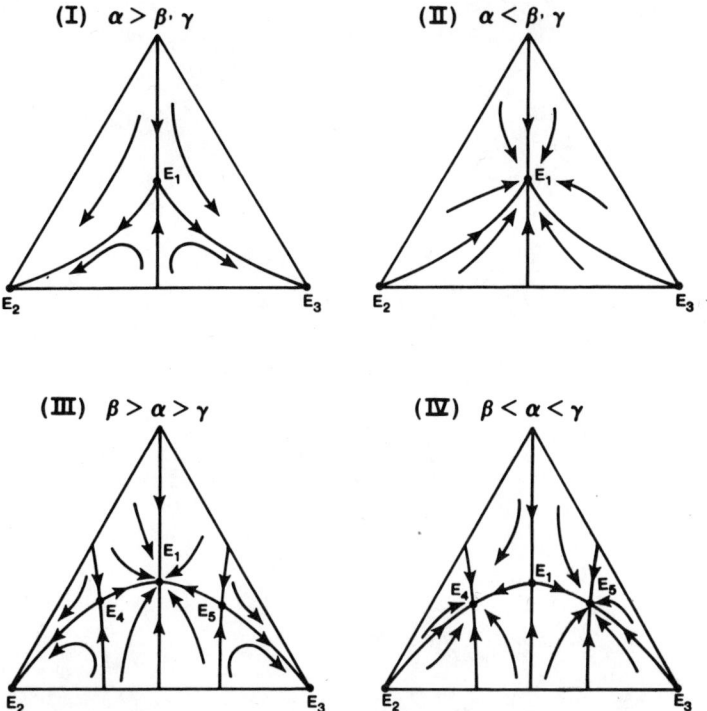

FIG. 4.1. *The vector field for the system* (4.3) *showing the four possible cases. From Butler, Freedman and Waltman* [11].

A novel view of Kamke's theorem is given by Hirsch [31]. Partially order the plane (componentwise ordering). Solutions which are noncomparable at time zero must remain noncomparable for all $t > 0$. This excludes limit cycles for any planar system (4.16) which satisfies (4.17), since every limit cycle in the plane has a critical point inside.

The original $(x_1(t), x_2(t), x_3(t))$ was assumed to be a nontrivial periodic solution of (4.2), so it transforms into periodic functions $\xi(t), \eta(t)$ which are among the solutions of (4.15)—a contradiction.

The Poincaré–Bendixson theorem then proves Theorem 4.1. The four possible cases are shown in Fig. 4.1 (from [11]).

The argument which excludes limit cycles is more general and allows more general (i.e., less symmetric) models to be considered [27]. However, the equations governing the location of critical points appear intractable.

References

[1] F. ALBRECHT, H. GATZKE, A. HADDAD AND N. WAX, *The dynamics of two interacting populations*, J. Math. Anal. Appl., 46 (1974), pp. 658–670.
[2] C. M. ANDERSON, J. ALLAN AND P. G. JOHANSEN, *Comments on the possible existence and nature of a heterozygote advantage in cystic fibrosis*, Modern Problems in Pediatrics, 10 (1967), pp. 381–387.
[3] A. A. ANDRONOV, E. A. LEONTOVICH, I. I. GORDON AND A. G. MAIER, *Qualitative Theory of Second Order Dynamical Systems*, John Wiley, New York, 1973.
[4] F. J. AYALA, M. E. GILPIN AND J. G. EHRENFELD, *Competition between species: Theoretical models and experimental tests*, Theoret. Population Biol., 4 (1973), pp. 331–356.
[5] K. BECK, *A model of the population genetics of cystic fibrosis in the United States*, Math. Biosci., 58 (1982), pp. 243–257.
[6] K. BECK, J. P. KEENER AND P. RICCIARDI, *Influence of infectious disease on the growth of a population with three genotypes*, preprint, 1982.
[7] ———, *The effect of epidemics on genetic evolution*, J. Math. Biol., to appear.
[8] G. S. BUTLER, *Coexistence in predator-prey systems*, in Modeling and Differential Equations in Biology, T. Burton, ed., Marcel Dekker, New York, 1980.
[9] G. J. BUTLER, S. B. HSU, AND P. WALTMAN, *Coexistence of competing predators in a chemostat*, J. Math. Biol., 17 (1983), pp. 133–152.
[10] G. J. BUTLER AND P. WALTMAN, *Bifurcation from a limit cycle in a two prey one predator ecosystem modeled on a chemostat*, J. Math. Biol., 12(1981), pp. 295–310.
[11] G. J. BUTLER, H. I. FREEDMAN AND P. WALTMAN, *Global dynamics of a selection model for the growth of a population with genotypic fertility differences*, J. Math. Biol., 14 (1982), pp. 25–35.
[12] R. P. CANALE, *Prey relationships in a model for activated process*, Biotech. and Bioengrg., 11 (1969), pp. 887–907.
[13] ———, *An analysis of models describing predator prey interaction*, Biotech. and Bioengrg., 12 (1970), pp. 353–378.
[14] K. S. CHENG, *Uniqueness of a limit cycle for a predator-prey system*, SIAM J. Math. Anal., 12 (1981), pp. 541–548.
[15] F. B. CHRISTIANSEN AND T. M. FENCHEL, *Theories of Populations in Biological Communities*, Springer-Verlag, Heidelberg, 1977.
[16] E. A. CODDINGTON AND N. LEVINSON, *Theory of Ordinary Differential Equations*, McGraw-Hill, New York, 1955.
[17] P. M. CONNEALLY, A. D. MERRITT AND P. YU, *Cystic fibrosis: Population genetics*, Texas Rep. Biol. Medicine, 31 (1973), pp. 639–750.
[18] W. A. COPPEL, *Stability and Asymptotic Behavior of Differential Equations*, D.C. Heath, New York, 1965.
[19] D. ERLE, *Stable closed orbits in plane autonomous dynamical systems*, J. Reine Angew. Math., 305 (1979), pp. 136–139.
[20] ———, *Nonuniqueness of stable limit cycles in a class of enzyme catalyzed reactions*, J. Math. Anal. Appl., 82 (1981), pp. 386–391.
[21] A. G. FREDRICKSON AND G. STEPHANOPOULOS, *Microbial competition*, Science, 213 (1981), pp. 972–979.
[22] ———, *Interactions of microbial populations in mixed culture situations*, preprint, 1982.
[23] H. I. FREEDMAN, *Deterministic Mathematical Models in Population Ecology*, Marcel Dekker, New York, 1980.

REFERENCES

[24] H. I. FREEDMAN AND P. WALTMAN, *Predator influence on the growth of a population with three genotypes*, J. Math. Biol., 6 (1978), pp. 367–374.
[25] G. F. GAUSE, *Vérifications expérimentales de la théorie mathématique de la lutte pour la vie*, Hermann, Paris, 1935.
[26] M. E. GILPIN AND F. J. AYALA, *Global models of growth and competition*, Proc. Nat. Acad. Sci., 70 (1973), pp. 3590–3593.
[27] K. P. HADELER AND D. GLAS, *Convergence to equilibrium in a population genetic model*, preprint, 1982.
[28] K. P. HADELER AND U. LIBERMAN, *Selection models with fertility differences*, J. Math. Biol., 2 (1975), pp. 19–33.
[29] J. K. HALE AND A. S. SOMOLINOS, *Competition for fluctuating nutrient*, preprint, 1982.
[30] S. R. HANSEN AND S. P. HUBBELL, *Single nutrient microbial competition; agreement between experimental and theoretical forecast outcomes*, Science, 20 (1980), pp. 1491–1493.
[31] M. HIRSCH, *Systems of differential equations which are competitive or cooperative* I: *Limit sets*, SIAM J. Appl. Math., 13 (1982), pp. 167–179.
[32] M. HIRSCH AND S. SMALE, *Differential Equations, Dynamical Systems, and Linear Algebra*, Academic Press, New York, 1974.
[33] C. S. HOLLING, *Some characteristics of simple types of predation and parasitism*, Can. Ent., 91 (1959), pp. 385–395.
[34] F. C. HOPPENSTEADT, *Mathematical Methods of Population Biology*, Cambridge Univ. Press, Cambridge, 1982.
[35] S. B. HSU, *Limiting behavior for competing species*, SIAM J. Appl. Math., 34 (1978), pp. 760–763.
[36] ———, *A competition model for a seasonally fluctuating nutrient*, J. Math. Biol., 9 (1980), pp. 115–132.
[37] S. B. HSU, S. P. HUBBELL AND P. WALTMAN, *A mathematical theory for single nutrient competition in continuous cultures of microorganisms*, SIAM J. Appl. Math., 32 (1977), pp. 366–383.
[38] ———, *A contribution to the theory of competing predators*, Ecol. Monogr., 48 (1978), pp. 337–349.
[39] ———, *Competing predators*, SIAM J. Appl. Math., 35 (1978), pp. 617–625.
[40] G. E. HUTCHINSON, *An Introduction to Population Ecology*, Yale Univ. Press, New Haven, CT, 1978.
[41] H. W. JANNASH AND R. T. MATELES, *Experimental bacterial ecology studied in continuous culture*, Adv. Microbial Physiol., 11 (1974), pp. 165–212.
[42] J. L. JOST, S. F. DRAKE, A. G. FREDRICKSON AND M. TSUCHIYA, *Interaction of Tetrahymena pyriformis, Escherichia coli, Azobacter vinelandii and glucose in a minimal medium*, J. Bacteriol., 113 (1976), pp. 834–840.
[43] J. P. KEENER, *Oscillatory coexistence in the chemostat: a codimension-two unfolding*, SIAM J. Appl. Math. 43 (1983), pp. 1005–1019.
[44] ———, *Oscillatory coexistence in a food chain model with competing predators*, J. Math. Biol., to appear.
[45] A. L. KOCH, *Competitive coexistence of two predators utilizing the same prey under constant environmental conditions*, J. Theoret. Biol., 44 (1974), pp. 378–386.
[46] J. P. LASALLE, *The Stability of Dynamical Systems*, CBMS Regional Conference Series in Applied Mathematics 25, Society for Industrial and Applied Mathematics, Philadelphia, 1976.
[47] J. E. MARSDEN AND M. MCCRACKEN, *The Hopf Bifurcation and Its Applications*, Springer, New York, 1976.
[48] R. M. MAY, *Stability and Complexity in Model Ecosystems*, Princeton Univ. Press, Princeton, NJ, 1973.
[49] J. MAYNARD SMITH, *Models in Ecology*, Cambridge Univ. Press, Cambridge, 1982.
[50] R. MCGEHEE AND R. A. ARMSTRONG, *Some mathematical problems concerning the ecological principle of competitive exclusion*, J. Differential Equations, 23 (1977), pp. 30–52.
[51] J. MONOD, *Recherches sur la croissance des cultures bactériennes*, Hermann, Paris, 1942.

REFERENCES

[52] T. NAGYLAKI AND J. F. CROW, *Continuous selection models,* J. Theoret. Population Biol., 5 (1974), pp. 257–283.

[53] E. O. POWELL, *Criteria for the growth of contaminants and mutants in continuous culture,* J. Genet. Microbiol., 18 (1958), pp. 259–268.

[54] E. B. PIKE AND C. R. CUIDS, *The microbial ecology of activated sludge process,* in Microbial Aspects of Pollution, G. Sykes and F. A. Skinner, eds., Academic Press, New York, 1971.

[55] G. SELL, *What is a dynamical system?,* in Studies in Ordinary Differential Equations, J. Hale, ed., Studies in Mathematics 14, Mathematical Association of America, Washington, DC, 1977.

[56] H. L. SMITH, *Competitive coexistence in an oscillating chemostat,* SIAM J. Appl. Math., 40 (1981), pp. 498–522.

[57] ———, *The interaction of steady state and Hopf bifurcation in a two-predator-one-prey competition model,* SIAM J. Appl. Math., 42 (1982), pp. 27–43.

[58] G. STEPHANOPOULOS, A. G. FREDRICKSON AND A. ARIS, *The growth of competing microbial populations in a CSTR with periodically varying inputs,* AIChE J., 25 (1979), pp. 863–872.

[59] F. M. STEWARD AND B. R. LEVIN, *Partitioning of resources and the outcome of interspecific competition; a model and some general considerations,* Amer. Naturalist, 107 (1973), pp. 171–198.

[60] H. VELDCAMP, *Ecological studies with the chemostat,* Adv. Microbial Ecol., 1 (1977), pp. 59–95.

[61] P. WALTMAN, S. P. HUBBELL AND S. B. HSU, *Theoretical and experimental investigations of microbial competition in continuous culture,* in Modeling and Differential Equations in Biology, T. Burton, ed., Marcel Dekker, New York, 1980.